D1756728

Handbook of Animal Science

Edited by

Paul A. Putnam
U.S. Department of Agriculture
Agricultural Research Service
Mid South Area
Stoneville, Mississippi

ACADEMIC PRESS, INC.
Harcourt Brace Jovanovich, Publishers
San Diego   New York   Boston
London   Sydney   Tokyo   Toronto

W 636 HAN

This book is printed on acid-free paper. ∞

Copyright © 1991 by ACADEMIC PRESS, INC.
All Rights Reserved.
No part of this publication may be reproduced or transmitted in any form or
by any means, electronic or mechanical, including photocopy, recording, or
any information storage and retrieval system, without permission in writing
from the publisher.

Academic Press, Inc.
San Diego, California 92101

United Kingdom Edition published by
Academic Press Limited
24–28 Oval Road, London NW1 7DX

Library of Congress Cataloging-in-Publication Data

Handbook of animal science / edited by Paul A. Putnam.
        p.   cm.
    Includes bibliographical references and index.
    ISBN 0-12-568300-6
    1. Livestock--Handbooks, manuals, etc.   2. Animal products-
-Handbooks, manuals, etc.   3. Animal industry--Handbooks, manuals,
etc.   I. Putnam, Paul A. (Paul Adin), date.
    SF65.2.H36   1991
    636--dc20   90-23933
                                                             CIP

PRINTED IN THE UNITED STATES OF AMERICA
91 92 93 94   9 8 7 6 5 4 3 2 1

Handbook of
Animal Science

Advisory Board

Timothy H. Blosser
S. 5516 Glendora Drive
Spokane, WA 99223
U.S. Department of Agriculture, ARS (Retired)

William P. Flatt
Dean and Coordinator
College of Agriculture
University of Georgia
101 Conner Hall
Athens, GA 30602

John K. Loosli
3745 NW 7th Place
Gainesville, FL 32607
Cornell University (Retired)

Everette J. Warick
4212 Van Buren Street
Hyattsville, MD 20782
U.S. Department of Agriculture (Retired)

Milton B. Wise
Vice President
Agriculture and Natural Resources
Clemson University
101 Barre Hall
Clemson, SC 29634

Handbook of Animal Science

Edited by

Paul A. Putnam

U.S. Department of Agriculture
Agricultural Research Service
Mid South Area
Stoneville, Mississippi

ACADEMIC PRESS, INC.
Harcourt Brace Jovanovich, Publishers
San Diego   New York   Boston
London   Sydney   Tokyo   Toronto

W 636 HAN

This book is printed on acid-free paper. ∞

Copyright © 1991 by ACADEMIC PRESS, INC.
All Rights Reserved.
No part of this publication may be reproduced or transmitted in any form or
by any means, electronic or mechanical, including photocopy, recording, or
any information storage and retrieval system, without permission in writing
from the publisher.

Academic Press, Inc.
San Diego, California 92101

United Kingdom Edition published by
Academic Press Limited
24–28 Oval Road, London NW1 7DX

Library of Congress Cataloging-in-Publication Data

Handbook of animal science / edited by Paul A. Putnam.
    p.   cm.
  Includes bibliographical references and index.
  ISBN 0-12-568300-6
    1. Livestock--Handbooks, manuals, etc.   2. Animal products-
  -Handbooks, manuals, etc.   3. Animal industry--Handbooks, manuals,
  etc.   I. Putnam, Paul A. (Paul Adin), date.
  SF65.2.H36   1991
  636--dc20   90-23933
                                                        CIP

PRINTED IN THE UNITED STATES OF AMERICA
91 92 93 94   9 8 7 6 5 4 3 2 1

Contents

PRODUCTION

PRODUCT/UTILIZATION

Contributors

Numbers in parentheses indicate the pages on which the authors' contributions begin.

Timothy H. Blosser (61, 385)
South 5516 Glendora Drive
Spokane, Washington 99223
U.S. Department of Agriculture
Agricultural Research Service
 (Retired)

B. C. Breidenstein (321)
8230 Springbrook Drive
Oklahoma City, Oklahoma 73132
Director of Research
National Livestock and Meat
 Board (Retired)
Chicago, Illinois 60611

Larry V. Cundiff (161)
U.S. Department of Agriculture
Agricultural Research Service
Meat Animal Research Center
Genetics and Breeding Research
 Unit
Clay Center, Nebraska 68933

Michael J. Darre (183)
Assistant Professor of Animal
 Science, Poultry
Department of Animal Science
University of Connecticut
Storrs, Connecticut 06268

Keith E. Gregory (161)
U.S. Department of Agriculture
Agricultural Research Service
Meat Animal Research Center
Genetics and Breeding Research
 Unit
Clay Center, Nebraska 68933

Andrew C. Hammond (271)
U.S. Department of Agriculture
Agricultural Research Service
Subtropical Agricultural Research
 Center
Brooksville, Florida 34605

H. W. Hawk (243)
U.S. Department of Agriculture
Agricultural Research Service
Beltsville Agricultural Research
 Center
Reproduction Laboratory
Beltsville, Maryland 20705

H. Herlich (203)
2907 Fallstaff Road, Apartment 41
Baltimore, Maryland 21209
U.S. Department of Agriculture
Agriculture Research Service
 (Retired)

Donald M. Kinsman (183)
Department of Animal Science
University of Connecticut
Storrs, Connecticut 06268

Robert M. Koch (161)
University of Nebraska
Meat Animal Research Center
Genetics and Breeding Research
  Unit
Clay Center, Nebraska 68933

Anthony W. Kotula (295)
U.S. Department of Agriculture
Agricultural Research Service
Beltsville Agricultural Research
  Center
Meat Science Research Laboratory
Beltsville, Maryland 20705

J. K. Loosli (25)
Professor of Animal Nutrition,
  Emeritus
Cornell University
Ithaca, New York

R. H. Miller (151)
U.S. Department of Agriculture
Agricultural Research Service
Beltsville Agricultural Research
  Center
Milk Secretion and Mastitis
  Laboratory
Beltsville, Maryland 20705

Paul A. Putnam (107, 173, 373)
U.S. Department of Agriculture
Agricultural Research Service
Mid South Area
Stoneville, Mississippi 38776

M. D. Ruff (203)
U.S. Department of Agriculture

Agricultural Research Service
Beltsville Agricultural Research
  Center
Beltsville, Maryland 20705

Theron S. Rumsey (261)
U.S. Department of Agriculture
Agricultural Research Service
Beltsville Agricultural Research
  Center
Ruminant Nutrition Laboratory
Beltsville, Maryland 20705

L. W. Smith (179, 385)
U.S. Department of Agriculture
Agricultural Research Service
Beltsville Agricultural Research
  Center
Beltsville, Maryland 20705

O. H. V. Stalheim (223)
1918 George Allen
Ames, Iowa 50010
U.S. Department of Agriculture
Agricultural Research Service
  (Retired)

Stephen A. Sulik (183)
Department of Animal Science
University of Connecticut
Storrs, Connecticut 06268

R. L. Willham (3)
Department of Animal Science
Iowa State University
Ames, Iowa 50010

J. C. Williams (321)
National Dairy Promotion and
  Research Board
2111 Wilson Blvd.
Ste. 600
Arlington, Virginia 22201

Preface

Animals and animal products have been a part of the human environment and heritage since prehistoric times. Animals provide a unique source of foods that have a high nutrient density and palatability. Animals are produced and utilized (for companionship, power, food) in all societies and geographic locations.

This publication presents information ranging from our livestock heritage, the history of animal nutrition, and a listing of animal breeds and populations to production characteristics, product definition, and consumption.

It is recognized that no publication can adequately cover all phases of animal science. Therefore, it was the intention of the editor to bring information together that is not generally available in a single text. References are provided that will lead the reader to specialized subject areas.

The authors were encouraged to limit prose to the extent necessary to facilitate use of the tabular information. The format is designed to simulate the classic handbooks of such sciences as chemistry and physics and complement the many specialized animal science texts.

A reference book of this type requires the cooperation and contributions of many specialists. I gratefully acknowledge the advisory board and the contributing authors for their submissions and patience.

Readers are encouraged to offer suggestions for changes and additions in future editions.

The editor wishes to express his appreciation to Flo Toffon, Leah Oppedal, Pat Brown, and Sandra Warren for their patience, persistence, and support as the manuscript was typed, edited, and retyped.

P. A. Putnam

History and Background

Our Livestock Heritage

R. L. Willham

Iowa State University of Science and Technology
Ames, Iowa

Introduction

The following time lines and illustrations are excerpted from R. L. Willham, "The Legacy of the Stockman" (Department of Animal Science, Iowa State University, Ames, Iowa, 1985). They capsulize the joint evolution of man and the animals he domesticated for food, power, transportation, religion, recreation, war, and companionship. The lines become shorter in time elapsed as they progress (e.g., 10 million years elapse in the Hunter, only 30 years in Scientific), reflecting the increased intensity and sophistication of man's manipulation of his livestock.

Handbook of Animal Science   Copyright © 1991 by Academic Press, Inc.
All rights of reproduction in any form reserved.

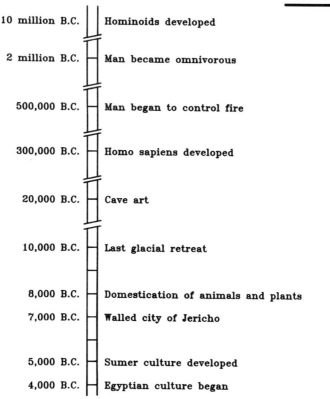

10 million B.C.	Hominoids developed
2 million B.C.	Man became omnivorous
500,000 B.C.	Man began to control fire
300,000 B.C.	Homo sapiens developed
20,000 B.C.	Cave art
10,000 B.C.	Last glacial retreat
8,000 B.C.	Domestication of animals and plants
7,000 B.C.	Walled city of Jericho
5,000 B.C.	Sumer culture developed
4,000 B.C.	Egyptian culture began

Note: Drawings from the period of history 15,000 B.C.–13,000 B.C. were found in the Hall of Bulls at Lascaux, Dordogne, France. The spirited specimens painted and carved on cave walls and ceilings tell of the precarious place of man. Charcoal and earth colors were used. The animals may have been created simply for pleasure, as symbols of the process of nature, as decoration for sanctuaries where religious rituals were performed, or for hunting rites to gain power over the true quarry. The caveman's artistry suggests the importance of the bull. Change: Man increased cereal production after he developed the scratch plow, yoked oxen to it, and began irrigation.

Drawing from the Hall of Bulls at Lascaux (15,000 B.C.–13,000 B.C.), Dordogne, France. The spirited specimens painted and carved on cave walls and ceilings tell of the precarious place of man. Charcoal and earth colors were used. The animals may have been created simply for pleasure, as symbols of the process of nature, as decoration for sanctuaries where religious rituals were performed, or for hunting rites to gain power over the quarry. The caveman's artistry suggests the importance of the bull.

3000 B.C.	Egypt--herdsmen attempted domestication of animals
2700 B.C.	China--swine household scavengers
2500 B.C.	Crete--bull prominent in culture Indus Valley--poultry and zebu cattle
2100 B.C.	Aryan invasion of India--sacred cow culture started
1730 B.C.	Hyksos invasion of Egypt--chariot horses brought to culture
1400 B.C.	Assyria--horses for hunting, cavalry, chariots
1200 B.C.	Mediterranean Basin--Phoenician horse traders
776 B.C.	Olympic games--chariot and horse races
753 B.C.	Rome--Equestrian Order began
650 B.C.	Settled cultures raided by mounted bowmen and pastoralists from steppes
560 B.C.	Persia--horse-borne communication network

Note: There were four river civilizations: Egypt on the Nile, Sumer on the Tigris and Euphrates in Mesopotamia, India on the Indus, and China on the Yellow.

The four river civilizations—Egypt on the Nile, Sumer on the Tigris and Euphrates in Mesopotamia, India on the Indus, and China on the Yellow.

Changes: Oxen were power source: sheep and goats provided clothing, meat, and fat; swine as scavengers gave meat and fat; poultry produced eggs and sport; and cattle and sheep were used as sacrificial animals.

TRADERS

490 B.C.	Greeks, without cavalry, defeated Persians at Marathon
430 B.C.	Greeks became traders after wars decimated agriculture
334 B.C.	Alexander the Great used cavalry in conquest of world
218 B.C.	Hannibal crossed Alps with zebu cattle
214 B.C.	Great Wall of China built to protect Chinese from mounted Mongols
5 B.C.	Birth of Christ
101 A.D.	Mithraic religion spread through Roman Empire Chariot races--Circus Maximus, Rome
313 A.D.	Christianity became state religion under Constantine
	Mounted Goths sacked Rome
475 A.D.	Successive waves of Eurasian horsemen ended Roman Empire

Note: Greeks wrote *Hornebook* on farm management—used by monks of Dark Ages.

Tradition: Oxen were power source; sheep and goats were meat and wool source; swine and poultry were raised on villas of the Basin; sheep and cattle still were sacrificial animals.

The Mediterranean Basin.

The Caledonian boar hunt in Greece.

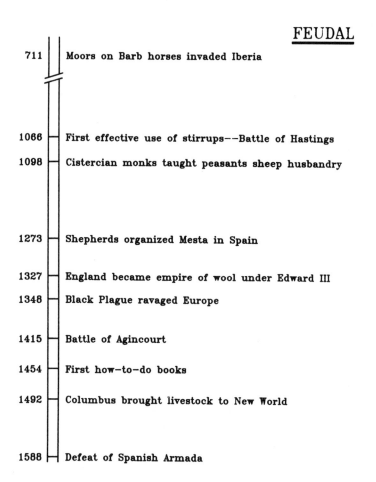

FEUDAL

711	Moors on Barb horses invaded Iberia
1066	First effective use of stirrups––Battle of Hastings
1098	Cistercian monks taught peasants sheep husbandry
1273	Shepherds organized Mesta in Spain
1327	England became empire of wool under Edward III
1348	Black Plague ravaged Europe
1415	Battle of Agincourt
1454	First how–to–do books
1492	Columbus brought livestock to New World
1588	Defeat of Spanish Armada

Tradition: Agriculture in Europe was manorial with livestock in commons. After 800 A.D., horses began to replace oxen because of use of horse collar. Swine became the garbage collectors in growing cities.

Cities of Importance to Europe during the Middle Ages.

Cistercian monk with the flock.

Drawing from the Bayeux Tapestry depicting the battle of Hastings.

COLONIAL

1521	Cortez used horses to conquer Tenochtitian
1540	Coronado accompanied by swine, sheep, and cattle
1550	Plains Indians add horse to culture
1598	Spanish cattle moved across Rio Grande to Texas
1607	Jamestown colonists forced to eat own livestock
1642	English civil war allowed colonists to develop livestock industry
1650	Colonies exported livestock
1776	Declaration of Independence
1783	Shorthorn cattle imported
1793	Invention of cotton gin

Tradition: British husbandry practiced in the East. Spanish feral husbandry developed in the Southwest. When the two systems joined, they produced a system unique to the Americas.

Conquistador's Sunbelt

British Colonies

PIONEER

1800	Pioneers crossed Cumberland Gap with livestock
1801	Agricultural societies promoted good animal husbandry
1803	Louisiana Purchase
1820	Missouri Compromise held agriculturally diverse nation together
1827	Erie Canal connected agricultural production with markets
1830	"Iron Horse" linked the nation
1837	John Deere plow opened prairie for farming
1849	California Gold Rush
1855	Michigan ag college pioneered agricultural education
1860	Civil War began

Change: Many livestock breeds were imported into the United States. Center of livestock production moved westward. Livestock production in east was intensified.

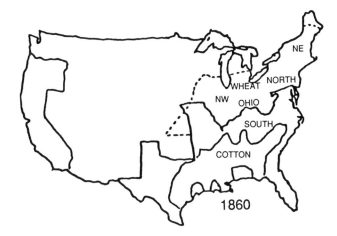

The United States at the start of the Civil War.

BRITAIN

1700s	Enclosure of manorial commons encouraged livestock breeding
1760	Bakewell set pattern for modern animal breeding
1769	Watt steam engine gave livestock larger market
1783	Bakewell organized Dishley Society, a breed association
1791	Thoroughbred studbook forerunner of herdbooks
1804	Napoleonic Wars increased demand for livestock products
1822	Coates's herdbook recorded Shorthorn ancestry
1840	Scotch Shorthorn developed for beef production
1874	Shorthorn society founded--other breed associations followed

Change: Many local strains of livestock were developed and a few became popular through promotion.

Robert Bakewell

England, Scotland, Ireland, and the French and Dutch coast of Europe where so many breeds of livestock developed.

EMPIRE

1862	Morill, Homestead, and USDA Acts revolutionized animal agriculture
1865	Chicago Union Stock Yards became hub of livestock industry
1867	Cattle drives began to railheads
1873	Barbed wire used on the Great Plains
1881	Breeder's Gazette first published
1884	USDA began livestock research (BAI)
1887	Hatch Act provided state agricultural experiment stations
1890	Babcock test for milk fat
1895	Trap nesting improved poultry production
1897	National Livestock Association organized

Change: Industry and agriculture developed in an empirical style.

Drawing of cowboy heeling a calf at roundup time.

Drawing of the gate at the Union Stock Yard of Chicago

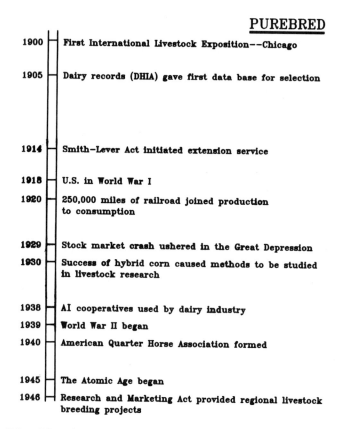

PUREBRED

1900 — First International Livestock Exposition—Chicago

1905 — Dairy records (DHIA) gave first data base for selection

1914 — Smith–Lever Act initiated extension service

1918 — U.S. in World War I
1920 — 250,000 miles of railroad joined production to consumption

1929 — Stock market crash ushered in the Great Depression
1930 — Success of hybrid corn caused methods to be studied in livestock research

1938 — AI cooperatives used by dairy industry
1939 — World War II began
1940 — American Quarter Horse Association formed

1945 — The Atomic Age began
1946 — Research and Marketing Act provided regional livestock breeding projects

Tradition: Most farm animal species were evaluated in the show ring.

One of the several bronze bulls used as logos for the International Livestock
Exposition in Chicago.
Change: The use of purebred sires was promoted.

22   R. L. Willham

SCIENTIFIC

1952	Freezing of bull semen from superior sires
1953	Gene chemical structure discovered – DNA
1954	Performance Registry International began
1955	Layer and broiler confinement industries separated
1957	Sputnik triggered increased interest in science and livestock research
1962	Milk production began dramatic increase through sire evaluation
1965	Computer technology revolutionized livestock data analysis
1967	Many beef breeds imported through Canada
1968	Beef Improvement Federation formed
1975	National Swine Improvement Federation organized
1980	Beef sire evaluation demonstrated genetic trend for growth

Change: The discovery of the structure of DNA in 1953 opened the door to new approaches to genetics, and science became critical to the animal industry.

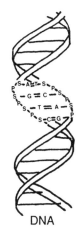

DNA

Structure discovered in 1953.

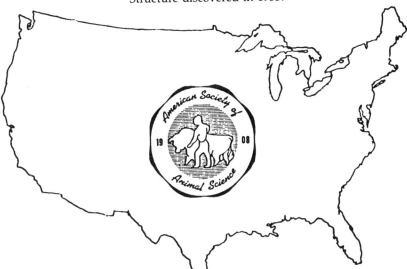

The logo of the American Society of Animal Science.

History of the
Development of Animal
Nutrition

J. K. Loosli

Department of Animal Science
Cornell University

Handbook of Animal Science   Copyright © 1991 by Academic Press, Inc.
All rights of reproduction in any form reserved.

Introduction

The development of the science of nutrition was one of the greatest advances in human history. Before the knowledge became available, millions of people in every generation led lives blighted by malnutrition. Suffering and inferiority of domestic animals, with consequent economic loss, was even more widespread throughout the world. Chickens could not lay eggs during the winter, and reproduction, growth, and livability of young pigs was largely limited to spring and summer seasons, when grass and sunshine abounded. There is no doubt that many useful facts have been discovered and lost because they were not recorded. Even since writing and printing became common, many facts have been rediscovered by people who did not read the literature of earlier discoveries.

In the temperate regions of the world, reproduction in farm animals was largely seasonal. The young were born in the spring, when the climate and feed supplies were most favorable. Advances in the application of nutritional knowledge and control of environmental factors now makes year-round production possible with domesticated animals. In the tropics, animal production has been controlled more by dry and rainy seasons than by other factors.

The meaning of the word "animal" includes all living organisms other than plants, from single-cell animals to humans. In keeping with current usage, however, in this chapter animal nutrition will be primarily concerned with the nutrition of farm animals, laboratory animals, and companion animals and not with humans, specifically. The nutrition of farm animals developed more rapidly in the late nineteenth and early twentieth centuries than did human nutrition, even though many early advances were made by physicians, physiologists, and chemists.

From the beginning of animal life, the procurement of sufficient food and water have been of primary concern. The Bible (Gen. 1:29, 30) records, "I have given you every herb. . . , the fruit of a tree, it shall be for meat, and every beast . . . and every fowl . . . I have given for meat." Early records of the importance of animals in the food supplies came to us from Egypt (Darby *et al.*, 1977). The Bible makes many references to animals as a source of food. Clearly, the early nomads depended greatly upon their animals for their food.

Contributions of Chemistry to Animal Nutrition

Scientific advances in nutrition depended upon developments in chemistry, especially in agricultural, biological, and physiological chemistry, as well as in physiology, pharmacology, medicine, and related sciences. In the 1600s, most development took place in England with Boyle, Harvey, Hook, Mayow, and others (Lusk, 1933). Boerhaave (1668–1738), the Dutch physician, published his book *Elements in Chemistry* in 1732. McCollum (1957) cites Rouelle (1718–1778) as the father of physiological chemistry. Men who participated in the chemical revolution, including Black, Cavendish, Lavoisier, Priestly, Rouelle, Scheele, and Spallanzani, were all born between 1718 and 1743. The men who discovered the essential mineral elements between 1669 and 1824 are listed in Table 1.

By the early 1800s, most of the essential elements had been discovered. Scientists in France, England, Sweden, and Holland had made most of the advances. France became the great center for learning during the last half of the 1700s, and young scientists from England and Germany went to study with Berthollet, Gay-Lussac, Laplace, Magendie, and Chevreul. The French Academie des Sciences in Paris was started in 1660. The Royal Society in Oxford, England, began in 1645 and was chartered in 1662.

The first scientific society in Germany began in 1650. After Liebig studied in France under Gay-Lussac, he became professor at Giessen and later Munich. With his students and associates—including Voit, Zuntz, Rubner, and others—Germany became the main center of learning for chemistry. Liebig organized the first chemical laboratory for students in Europe. He was the greatest organic chemist of his time (McCollum, 1957), he improved methods for analyzing feeds and tissues. His book *Thierchemie*, published in 1842, and Voit's book *Stoffwechsel und Ernahrung*, published in 1881, became classics. German workers made great advances in protein and energy nutrition (McCay, 1973). German scientists isolated 14 of the 23 amino acids (Table 2). Most of the early energy metabolism research, but none on vitamins, was done in Germany.

Table 1
Discovery of the Essential Mineral Elements

Element	Date	Person and Events
Iron	Biblical	Bible records Tubal-Cain was an instructor in iron and brass
	Before 1680	Sydenham; treated anemia with iron filings in wine
	1713	Lemery; found iron in blood
Zinc	Biblical	Bible reports brass (a mixture of zinc and copper)
	1300	Metallic zinc used
	1746	Marggraf rediscovered in Europe
	1922	Bertrand and Berzon; necessary for growth of rats
	1955	Tucker and Salmon; essential for pigs on normal feeds
Copper	Biblical	Bible reports use of brass, mined for more than 5000 years
	1300	Records of mining in India
	1928	Hart et al.; needed for hemoglobin formation in rats
Sulfur	Biblical	Bible refers to brimstone, crude sulfur
	1784	Scheele; reported sulfur in proteins
Tin	Known to ancients	
	1970	Schwarz; necessary for metallo-enzymes
Fluorine	1529	(Georgius) Agricola used fluorospar as flux
	late 1700s	Scheele et al. described hydrofluoric acid
	1886	Moisson; isolated element
	1972	Schwarz; needed for teeth and bones
Phosphorus	1669	Brand; isolated from urine
	1748	Gagn; present in bones
	1770	Scheele; bones contain calcium phosphate
Cobalt	1735	Brandt; first discovery
	1935	Underwood et al.; dietary essential for ruminants
Nickel	1751	Cronstedt; discovery
	1973	Anke; growth of pigs
	1974	Nielsen and Ollerich; essential for rumen bacteria
Calcium	1770	Scheele; bones contain calcium phosphate
	1842	Chossat; needed for bone development
Chlorine	1774	Scheele; discovery
	1810	Davy named the element
	1823	Prout; hydrochloric acid in gastric juice
Magnesium	1755	Black recognized the element
	1808	Davy isolated magnesium
	1924	Theiler et al.; essential for cattle
Manganese	1774	Gahn discovered
	1931	Kemerer and McCollum; essential for rats and mice
Molybdenum	1778	Scheele; discovered
	1872	Hjelm; prepared in pure form

(*continued*)

Table 1   (*Continued*)

Element	Date	Person and Events
	1938	Ferguson; toxicity shown for animals
	1953	Richert and Westerfield; metallo-enzymes xanthine oxidase
Chromium	1797	Vauquelin discovered
	1959	Schwarz; essential for glucose metabolism
Vanadium	1801	del Rio discovered, not accepted
	1830	Sefstrom rediscovered
	1971	Schwarz and Milne
	1975	Hopkins and Mohr; needed for oxidation-reduction catalyst
Potassium	1807	Davy; from caustic potash (KOH). First metal isolated by electrolysis
	1847	Liebig; present in animal tissues
Sodium	1807	Davy; long recognized in compounds (e.g., common salt)
	1905	Babcock; needed by lactating cows
Iodine	1811	Courtois discovered
	1825	Boussingault; stated iodine was the only specific cure for goiter
	1896	Bauman; thyroid rich in iodine
	1920	Coindet; iodine in burnt sponges explained its effectiveness. Thyroxine discovered.
Selenium	1817	Berzelius; in flue dust from copper smelters
	1934	Franke and Potter; toxic element in certain plants
	1957	Schwarz; necessary for rats
Silicon	1824	Berzelius
	1970	Carlisle; needed for bone calcification

Sources: McCollum, 1957; McCay, 1973; Maynard, 1937; Weast.

The young American scientists who studied in Germany under Liebig, Voit, Rubner, and Kellner included Armsby, Atwater, Lusk, Mendel, and E. B. Hart, who became the leaders in nutrition research. Russell H. Chittenden (1956–1943) in the late 1800s organized the first laboratory to teach physiological chemistry at Yale University, where many American nutritionists were trained. Gradually, the greatest center of learning shifted to America in the late 1800s and early 1900s, as the land-grant university and experiment station systems developed.

Discovery of the Essential Nutrients

Following is a condensed summary of the discoveries of the nutrients required by animals and of the persons who made the important discoveries. The facts are presented largely in tabular form with only a limited amount of supporting material in the text.

Table 2
Discovery of the Amino Acids

Element	Date	Person and Reference
Essential amino acids		
Leucine	1819	Proust and Braconnot, French; *Ann. Chim. Phys.* **13**, 113
Valine	1856	Gorup-Besanez, German; *Ann.* **98**, 1
Phenylalanine	1879	Schulze, German; *Ber.* **12**, 1924
Arginine	1886	Schulze, German; *Ber.* **19**, 1177
Lysine	1889	Dreschsel, German; *J. Prakt. Chem.* **39**, 425
Histidine	1896	Kossel-Hedin, German-Swedish; *Z. Phys. Chem.* **22**, 176–191
Tryptophane	1901	Hopkins and Cole, English; *J. Phys.* **27**, 418
Isoleucine	1903	Erlich, German; *Ber.* **37**, 1809
Methionine	1922	Muller, American; *Proc. Soc. Expt. Med.* **19**, 161
Threonine	1935	Rose, American; *J. Biol. Cyhem.* **112**, 283
Nonessential amino acids		
Cystine	1810	Wollaston, English; *Phil. Trans. Roy. Sco*, 223
Glycine	1820	Braconnot, French; *Ann. Chim. Phys.* **13**, 113
Aspartic acid	1827	Plisson, French; *J. Pharm.* **13**, 474
Tyrosine	1846	Liebig, German; *Ann.* **57**, 127; **69**, 116
Alanine	1850	Strecker, German; *Ann.* **75**, 27
Serine	1865	Cramer, German; *J. Prakt. Chem.* **96**, 76
Glutamic acid	1866	Ritthausen, German; *J. Prakt. Chem.* **99**, 454
Iodogogoic acid	1896	Dreschsel, German; *Z. Biol.* **33**, 96
Proline	1900	Willstatter, German; *Ber.* **33**, 1160
Hydroxyproline	1902	Fischer, German; *Ber.* **35**, 2660
Thyroxine	1915	Kendall, American; *J. Biol. Chem.* **39**, 125
Citrulline	1930	Wada, Japanese; *Proc. Imp. Acad. Japan* **6**, 15

It is now known that 47 to 50 specific nutrients are required in the diet of man and animals for normal health and performance. The essential nutrients include at least 20 mineral elements, 13 vitamins, 10 to 12 amino acids, and one fatty acid (linoleic). In addition to the above nutrients, animals need oxygen (O_2), supplied by the air they breathe, carbon (C) for carbohydrates, fats or protein, nitrogen (N) from protein, amino acids or certain other N-containing compounds, and sulfur (S) from protein or sulfur-amino acids. Next to oxygen, water is the most critical need of the body. Most mature animals can survive for weeks without food but only a few days without water. Energy, the greatest quantitative need, is provided by carbohydrates, fats, and protein. Protein or the essential amino acids are also required in a fairly large quantity.

It is not difficult to understand the many problems scientists faced in sorting out from the hundreds of chemical elements and thousands of organic compounds those 50 specific ones that are essential to life. The absence of any one of the essential nutrients will result in failure of growth of young animals, ill health, or even death at any age. While the minerals and vitamins are needed in only small or even trace amounts, they must be provided in the correct amount, since excesses of some of them will cause toxicity or death. The animal body has the ability to synthesize many organic compounds to build specific cells or tissues. It produces the enzymes and hormones needed to regulate metabolism, if its food supplies the necessary building blocks. Although there are major differences in the anatomy of the digestive tracts of various animals and birds and the kinds of foods they eat, there are only minor differences in the specific nutrients they require.

Discovery of the Essential Mineral Elements

There are no records that document the original discovery of at least five of the essential mineral elements, including iron, zinc, copper, sulfur, and tin. Common salt was an item of trade before recorded history that satisfied the salt cravings of grazing animals and was used to flavor foods. Table 1 lists the earliest records of the discoveries of the mineral elements now known to be dietary essentials. Only the earliest discovery of the element and the first demonstration of its essential nature are shown. Details of the discoveries and later developments are given by Lusk (1933), McCollum (1957), and McCay (1973). Information about the functions, metabolism, and requirements of the minerals are available in modern books on nutrition, such as Maynard and associates (1979).

Discovery of the Essential Organic Nutrients

The importance of water to sustain plant and animal life has been recognized from the earliest times. The difficulty of providing an adequate supply of pure, fresh water remains a key problem in many areas of the world. Next to the air we breathe, water is the most critical need for all forms of life on earth.

Hippocrates (460—370 B.C.), the father of medicine, initiated a long period of philosophizing about foods (Williams, 1978). He postulated a single universal nutrient substance present in various products that existed in several forms. This idea persisted even in William Beaumont's report on digestion studies with St. Martin in 1833 (McCollum, 1957). The *Great Guides to Science*, written by Francis Bacon (1561–1626), had contributed to the advances of scientific thought. It was not until about this period that the views of the Greek philosophers such as Aristotle and others that the main elements or principles were earth, air, water, and fire were questioned. In his extensive works, Robert Boyle (1627–1691) had postulated that life was a slow burning process. He recorded many other observations that advanced the knowledge of chemistry and nutrition. The discoveries of carbon dioxide (CO_2) by Black in 1757, of hydrogen (H_2) by Cavendish in 1766, of nitrogen (N) by Rutherford and by Scheele independently in 1772, of oxygen (O_2) by Priestly in 1774, and of the composition of water by Cavendish in 1783 and the development of animal calorimetry by Crawford in 1779 were all misunderstood until Lavoisier interpreted the knowledge in terms of the science of chemistry and nutrition and stated that "life is a chemical function." In the 1770s, he pointed out the chemical basis of respiration and metabolism and inaugurated the scientific basis of nutrition (Lusk, 1933).

Energy Metabolism

Based on the diets they consume, animals can be classified as herbivores (plant eaters), carnivores (eaters of animals and their products), or omnivores (eaters of every kind of food). Most of this food consumed is used to supply energy.

Lusk (1933) stated that Hippocrates attributed the development of medical knowledge to problems of nutrition. He wrote that animal heat was derived from fuel like a flame. He visualized that the body needed a single nutrient which was obtained from various foods. Erasistratus (310–250) was probably the first experimental physiologist. He devised a respiration apparatus by putting fowls in a jar and weighing them before and after they were fed. Crawford (1748–1795), a pupil of Black, is said to have been the first to measure animal heat. Lavoisier (with Laplace) determined the oxygen and carbonic acid involved in the respiration of a guinea pig in an ice calorimeter and explained the role of oxygen in heat production in 1780. In studies with a human in 1787, they found that the intensity of oxidation depended on food, environmental temperature, and mechanical work. In 1849, Regnault and Reiset studied animals in a bell jar, a closed circuit respiration apparatus.

Voit (1831–1908), the father of metabolic research, with Pettenkofer developed a respiration apparatus sufficiently large for a dog or a human. Max Rubner (1854–1932), working in Voit's laboratory, built the first accurate respiration calorimeter to measure heat production of dogs. He demonstrated that calories from fats and carbohydrates were of equal value but that protein had a lower metabolizable energy value when burned because of the energy in nitrogen end products. He formulated the surface area law, reported the specific dynamic action of foods, and set the stage for modern energy studies.

In 1842, J. P. Joule presented the foundations for the mechanical equivalent of heat. J. R. Meyer announced the law of the conservation of energy in 1845, which Rubner proved in 1891. The early American researchers in energy metabolism studied in Germany with Voit, Rubner, Kuhn, Zuntz, and others. These included Atwater, Armsby, Lusk, E. B. Hart, and others. Nathan Zuntz (1848–1920) built the first portable respiration apparatus, which was used extensively by his pupil Magnus-Levy with hospital patients. Atwater, with E. B. Rosa, constructed the first respiration calorimeter in the United States in 1892–1897 at the first U.S. agricultural experiment station, established at Middletown, Connecticut, in 1875. He reported the fuel value of many foods in 1899. F. G. Benedict joined the work and carried out energy metabolism studies with man and many animals for 40 years. Oscar Kellner in Germany conducted many respiration studies with farm animals and published his starch values for 300 feeds in 1905.

Armsby built a respiration calorimeter at the Pennsylvania experiment station and published energy balance data in 1903. He developed his net energy system of feed evaluation. This work was continued by E. B. Forbes, Max Kriss, R. W. Swift, and their associates. Lusk studied energy and protein metabolism in dogs and man at the Cornell Medical College and published his findings in 1932 (Lusk, 1932). Brody's book (1945) summarizes his extensive research on energy metabolism.

Most recent energy research has been carried out by Breirem in Norway, Blaxter (1962) in Scotland, and Flatt, Moe, and Tyrrell at the U.S. Department of Agriculture and their many associates (Maynard and Loosli, 1969).

Nitrogen, Protein, and Amino Acids

Ammonium salts were probably articles of commerce in Roman times. Geber described a method to make nitric acid in the thirteenth century. Boyle prepared impure ammonium carbonate by distilling animal products mixed with potash or lime. He was the first to collect nitrogen and nitric oxide gas. Berthollet was the first to recognize nitrogen as a constituent of animal tissues in 1791. He determined the ratio of hydrogen to nitrogen in ammonia. Ammonia as a gas had been recognized by Stephen Hales in 1727. Rutherford had isolated nitrogen gas in 1772. Fourcroy found nitrogen in the air bladder of carp. Before this, Rouelle had discovered urea and hippuric acid in urine, and Wohler produced it from ammonium cyanate in 1828. Nitrogen-containing compounds were not well understood until Fourcroy's book, published in 1800, described nitrous and nitric acids. Magendie in 1816 was first to demonstrate that body nitrogen arose from food nitrogen and that atmospheric nitrogen was not utilized by animals. His report on gelatin in 1816 made it clear that there were differences in proteins and that gelatin could not replace meat in the diet of dogs. Mulder (1802–1880) gave the name "protein" to the nitrogen-containing radical in various foods. Wollaston isolated the first amino acid, cystine, in 1810. Leucine and glycine, the first amino acids to be obtained by hydrolysis of protein, were identified in 1820. The last essential amino acid to be discovered was threonine, by Rose in 1935.

A summary of the discoveries of the amino acids is shown in Table 2. Vickery and Schmidt (1931) reviewed the early history of the discovery of the amino acids. Until all of the essential amino acids were known and available, it was not possible to make carefully defined purified diets that were needed to discover the vitamins and essential trace mineral elements (McCay, 1973).

Escher, in studies with dogs, found in 1850 that a deficiency of gelatin could be corrected by tyrosine. In 1857, Voit found, contrary to Liebig's view, that muscular work did not increase protein metabolism in a dog if adequate energy was available. Weiske showed in 1879 that nitrogen in asparagine was utilized by sheep and rabbits, but Voit found this was not true for dogs. That same year, E. Schule devised methods for measuring peptone, nitrate, and ammonia fractions in plants. Voltz found that urea could serve ruminants as a nitrogen source. Osborne and Mendel showed that diets containing certain proteins resulted in reduced growth when fed alone but were satisfactory when supplemented with the missing amino acids (Maynard and Loosli, 1969).

Full realization that proteins control chemical reactions in the body came after J. B. Sumner isolated and crystallized the first enzyme, urease, in 1926 and proved it was a protein. The world-renowned German chemists had as much difficulty accepting this fact as they did the vitamin theory earlier. It was not until 1937 that Sumner was awarded the Sheele Medal and 1946 the Nobel Prize in chemistry.

The Lipids, Essential Fatty Acids, and Choline

We owe much of our knowledge about the chemistry of lipids to the research of Chereul (1787–1889) in Paris (McCollum, 1957). Burr and Burr first demonstrated in 1929 with rats that linoleic acid was a dietary essential (Maynard, 1937). To this date, no other lipid has been found essential. This fatty acid is found most abundantly in corn, soybean, and certain other vegetable oils but not in coconut oil. As a group, the lipids (fats) supply more than twice as much energy per unit weight as do carbohydrates or proteins. The different animal and vegetable fats are largely interchangeable as an energy source, although there are some differences in digestibility. Lipids contribute to the acceptability of foods. They serve as a carrier of the fat-soluble vitamins.

In 1844, Gobley isolated a substance, which he named lecithin, from the lipids of egg yolk. This substance contained both nitrogen and phosphorus, along with fatty acids and glycerine. On hydrolysis, it yielded a nitrogen base which Strecker identified as choline in 1868 (McCollum, 1957). It was not until 1947 that choline was shown to be important for growth of young animals and birds. Choline must be supplied in the diet except to the extent that it can be synthesized within the body from labile methyl groups through transmethylation (Maynard *et al.*, 1979).

Discovery of the Vitamins

Some of the diseases now known to be caused by vitamin deficiencies date far back in history. Scurvy was described as early as 1500 B.C. Beriberi was apparently known to the Chinese by that time or earlier. Other maladies that afflicted people included rickets, night blindness, xerophthalmia, anemia, and pellagra. Recent analysis by Mar and Read of substances used very early by the Chinese as a remedy for night blindness showed they were richer in vitamin A than cod liver oil (Williams, 1978). Cod liver oil was used as a cure long before anything was known about the cause of rickets. Scurvy was a particular plague among groups on limited diets. Hippocrates described scurvy among soldiers in about 400 B.C. It caused death in four-fifths of Magellan's sailors and in 100 of 160 of Vasco da Gama's men. In 1536, Indians in Canada cured scurvy in Jacques Cartier's men with a broth of pine needles. Sir Richard Hawkings cured scurvy in the British Navy with citrus fruit. These experiences did not become widely known until much later and scurvy continued to occur (Table 5). In 1747, James Lind prevented scurvy with citrus, but its routine use was not started in the British navy until 1795.

In 1816, Magendie described xerophthalmia in dogs fed on pure carbohydrate, olive oil, and water in a study of dietary protein, but this remained hidden in the French literature until McCay (1973) cited the report. In 1881, Lunin, a student of Von Bunge, reported studies showing that mice would not grow on a mixture containing the proximate principles of milk, while they grew well on fresh milk. He suggested that milk contained unknown substances essential to life in addition to the three proximate principles and minerals. This report stimulated the search for the unknown factors. Takaki studied beriberi in the Japanese navy. The improvement resulting from a change in diet was attributed to extra protein. In the 1890s, Eijkman reported that a factor in rice bran prevented beriberi in humans and polyneuritis in pigeons. In 1905, Pekelharing, at the University of Utrecht, reported there was an unknown factor in milk necessary for mice. When he fed mice a baked diet of rice flour, egg albumin, casein, lard, salts, and water, all animals died in a few weeks, but with fresh milk or whey they remained healthy. In 1906, Hopkins wrote that scurvy and rickets were disorders caused by diets deficient in unidentified nutrients.

In 1910, McCollum pointed out the lack of knowledge on the relation of foods to health (McCollum, 1957). It generally was thought that a supply of protein, fat, carbohydrates, and an ill-defined supply of inorganic salts were all that animals required. Pasteur's (1822–1895) research showing that bacteria caused disease led medical doctors to doubt that certain diseases resulted from nutrient deficiencies in foods (i.e., the "vitamin theory"). Many observations relating specific foods to the cure or prevention of deficiency diseases finally led to the isolation of specific vitamins. This was possible only after knowledge became available about the chemistry of proteins, carbohydrates, fats, and some of the minerals, and their importance in the diet of animals.

The discovery of vitamins A and D was closely related to studies with cod liver oil. This oil has been used since very early times by the Greenlanders, Laplanders, and Eskimos. McCay (1973) summarized the early information about Moller's introduction of the steam process in 1853 to replace the rotting method for production of oil from cod livers. His studies appeared to indicate that the oil had general nutrition value but contained no specific agent in the small, nonsponifiable fraction. Some early physicians believed it had little value in treating rickets. Cod liver oil was fed to farm animals as early as 1824. Swine thrived on 60 ml per day, but a larger amount gave a fishy flavor to the meat. In 1854, an article in the *Farmers Magazine* described the yellow fat and injury to animals caused by feeding too much cod liver oil. In 1880, rickets in young lions in the London Zoo were cured by feeding them crushed bones and cod liver oil. The caloric value of cod liver oil was reported in 1866.

In 1912, Casimir Funk coined the term "vitamine" for the factor from rice bran containing nitrogen. This term became the common name for the group of organic substances required by animals in trace amounts. Eddie (1941) reviewed the details of the discovery of the vitamins.

Vitamin A   In 1913, McCollum and Davis, using a purified diet with rats, demonstrated that butter fat and egg yolk contained an essential fat-soluble substance. They called it fat-soluble A. Independently, Osborne and Mendel found at the same time that cod liver oil stimulated growth when added to a basal diet on which growth rate was subnormal. Evidence that there was an unknown fat-soluble factor as well as the water-soluble anti-beriberi substance, later called vitamin B, stimulated interest and research to identify the substances. The most important findings related to vitamin A are summarized in Table 3.

Table 3
Discovery of Vitamin A

Date	Persons and Events
1909	Hopkins reported a rat growth factor in some fats[a]
1913	McCollum and Davis; discovered fat-soluble A in butter
1913	Osborne and Mendel; discovered rat growth factor in cod liver oil
1919	Steenbock; reported the yellow color of vegetables was vitamin A
1929	Moore; animal body converts carotine into vitamin A
1935	Wald; relation of vitamin A to night blindness and vision
1942	Baxter and Robeson crystallized vitamin A
1947	Isler *et al.* synthesized vitamin A
1950	Karrer and Inhoffen synthesized carotine

[a]The Chinese used selected foods to cure night blindness centuries earlier. The earliest reference to vitamin A deficiency was in Jeremiah 14:6, "And the wild asses did stand in high places, their eyes did fail, because there was no grass."

McCollum (1957) reviewed the details of the early research. Steenbock's observation that the yellow color of vegetables was associated with vitamin A activity was not explained until Moore showed in 1929 that carotene was converted to vitamin A in the animal body. Studies in the 1920s and 1930s showed that most animal species needed a dietary source of vitamin A. The simultaneous use of chemical methods and experimental rats to test the products obtained resulted in the successful demonstration of the activity of vitamin A as the first vitamin, rather than vitamin B or C, which had received earlier attention. Similar testing methods were used to identify some other vitamins.

Vitamin B-1 (Thiamin)   Beriberi, one of the syndromes of vitamin B-1 deficiency, had been a problem for a long time. At least by 1896, it was realized that the disease was caused by feeding polished rice but not crude rice. Table 4 lists the main events that led to acceptance of the existence of the vitamin. Attempts to determine the vitamin B requirement of cattle led to the discovery that rumen bacteria synthesized the vitamin and that a dietary source is not required for ruminants, except under unusual circumstances.

Vitamin C   When long ocean voyages began, scurvy became a serious problem. In 1779, the Channel Fleet had 2400 cases of scurvy after a 10-week cruise. McCollum (1957) reviewed the problems in detail. Unlike beriberi, scurvy was seen only in humans and not in animals or birds.

Table 4
Discovery of Vitamin B-1 (Thiamin)

Date	Persons and Events
1896	Eijkman; a dietary factor in rice bran prevented beriberi
1912	Funk; suggested "vitamine" for the anti-beriberi substance
1928	Jansen and Donath isolated the factor in rice bran and found it contained C, H, O, and N; named it aneurine
1928	Bechdel *et al.*; rumen bacteria of cattle synthesized vitamin B
1936	R. R. Williams *et al.*; synthesized thiamin

The demonstration in 1907 by Holst and Frolich had scurvy could be produced in guinea pigs on the same foods that caused the disease in man provided an experimental animal for research. It was not until the 1930s that the chemical properties of vitamin C were described. Finally, it became evident that all primates required the vitamin in their diets. The main events leading to full understanding of the nature of the vitamin are listed in Table 5. McCay cited a physicians' book published in 1800 in which Willich suggested a diet of three parts fresh vegetables

Table 5
Discovery of Vitamin C (Ascorbic Acid)

Date	Persons and Events
1550 B.C.	Scurvy described in Ebers Papyrus (Phebes)
600 B.C.	Hippocrates described scurvy in soldiers
1492–1600	Magellan; four-fifths of his sailors died; Vasco da Gama; 100 of 160 men died
1536	Jacques Cartier; Canadian Indians cured scurvy with broth of evergreen needles
1564	Garrison stated the Dutch used lemon juice as a cure
1593	Sir Richard Hawkins; citrus cured scurvy in British navy
1747	John Huxham, a pupil of Boerhaave, recommended a vegetable diet for 1200 sailors who had scurvy
1753	James Lind; citrus prevented scurvy on shipboard
1795	British navy added lemon juice to ration
1883	Sir Thomas Barlow differentiated rickets from scurvy
1906	Hopkins suggested scurvy was a deficiency disease
1907	Holst and Frolich produced scurvy in guinea pigs
1928	Szent-Gyorgi isolated hexuronic acid (vitamin C)
1932	Waugh and King isolated hexuronic acid from lemon juice
1933	Harworth determined structure of ascorbic acid
1933	Reichstein synthesized scorbic acid (vitamin C)

and one part meat to prevent scurvy. Why did it take so long to accept and use this knowledge?

Vitamin D   Rickets and bone deformities in children had been a problem in some areas for centuries. Although cod liver oil had been used as a cure, this was not known generally among physicians. The key discoveries that led to a knowledge of the role of vitamin D in preventing rickets are summarized in Table 6. Mellanby thought vitamin A activity was involved with rickets. McCollum's oxidized cod liver oil, which destroyed the vitamin A activity, still cured rickets, proving that the two were different.

Vitamin E   A number of researchers reported poor reproduction in rats fed semipurified diets. In 1922, H. M. Evans and H. M. Bishop observed that wheat germ oil prevented resorption of embryos in pregnant females. Karl Mason (1900–1978) described a diet-induced degeneration of the germinal epithelium in male rats in 1925. The condition could be prevented but not cured. Evans isolated alpha- and beta-tocopherol from wheat germ oil. Other isomers were identified later.

Vitamin K   In 1929, Henrik Dam in Denmark noted hemorrhages in chicks fed a fat-free diet. In 1930, Horvath first mentioned that an unknown factor was necessary for blood clotting in chickens. In 1931, McFarlane *et al.* were the first to relate hemorrhages in chickens with the

Table 6
Discovery of Vitamin D

Date	Persons and Events
1650	F. Glisson, a professor at Oxford, characterized rickets; British physicians cured rickets in children with sunlight or ultraviolet light
1919	Mellanby produced rickets in dogs fed oatmeal; cod liver oil effected a cure
1922	McCollum *et al.*; oxidized cod liver oil cured rickets
1923	Goldblat and Soames; livers of irradiated rats contained antirachitic factor
1926	Steenbock; irradiation of foods as well as animals produced vitamin D
1926	Shipley *et al.* showed the relation of Ca and P to rickets
1931	Askew *et al.* determined the structure of ergosterol
1932	Windaus *et al.* determined the structure of 7-dehydrocholestrol (D_3)
1970	Deluca; 1,25-dihydroxy-vitamin D_3, active metabolite

Table 7
Discovery of Riboflavin

Date	Persons and Events
1926	Goldberger and Lillie described a rat syndrome, later shown to be riboflavin deficiency
1929	Norris *et al.* reported a curled-toe paralysis in chicks
1933	Kuhn and Gyorgy isolated a yellow pigment from egg white; suggested the name flavin
1935	Kuhn in Germany and Karrer in Switzerland synthesized riboflavin
1935	Gyorgy showed the synthetic compound had the same activity as the natural one

feeding of an ether-extracted fish meal. Hemorrhages were prevented when unextracted fish meal was included in the diet. Following further research, Dam named the substance vitamin K.

The research journals of Almquist and Stokstad revealed that they had discovered vitamins K-1 and K-2 in 1928 in studies with chicks. University administrators delayed the paper they had prepared and finally, when it was submitted to science, it was rejected. At that time, Dam's research was published. Doisey also contributed to the knowledge of the role of vitamin K in blood clotting (Jukes, 1980). Later when Dam (with Doisey) received the Nobel Prize for the discovery of vitamin K, it is reported that he had expected Almquist to share the prize.

For a time when McCollum discovered vitamin A, several scientists believed that there were only two vitamins, water-soluble B and fat-soluble A. In 1919, H. H. Mitchell published a critical review of the literature to show that water-soluble B was multiple in nature. This stimulated research on the B vitamins.

Riboflavin   Riboflavin was the second member of the vitamin B complex identified (Table 7). In 1879, Blyth reported a yellow pigment in milk whey. Bleyer and Kallman described the properties of the pigment in 1925. Warburg and Christian (1932) described an enzyme from yeast, which they called lumiflavin, that was yellow in solution and gave a green fluorescence. The yellow pigment turned out to be riboflavin.

Vitamin B-6 (Pyridoxine, Pyridoxal, Pyridoxamine)   In 1934, Gyorgy separated vitamin B-1 and riboflavin from the crude extract. This factor, which he named vitamin B-6, would cure dermatitis in young rats. Five other factors had been suggested earlier. In 1938, Lepskowsky isolated crystals of his earlier reported "factor I." Kuhn determined the structure of the factor and called it adermin. Gyorgy proposed the name, pyridoxine, after Harris and Folkers had synthesized the vitamin. Two other active forms were later identified.

Pantothenic Acid   In 1933, R. J. Williams and associates fractionated a growth factor from yeast and named it pantothenic acid. Other researchers found it was the same factor concerned with dermatitis in chicks and other animals.

Nicotinic Acid (Niacin)   Pellagra was known in Europe as early as 1795. It was first reported in the United States (New York and Massachusetts in 1864). In 1920, Goldberger reported that pellagra was not caused by bacterial infection but rather by an ill-balanced diet high in corn. He observed that pellagra in man resembled black tongue in dogs fed similar foods. In 1937, Elvehjem and associates found that nicotinic acid, a chemical that had been isolated from rice polishings by Funk in 1912, cured black tongue in dogs. It was also quickly shown to be effective in curing pellagra in humans.

Biotin   In 1926, Boas reported a dermatitis in rats fed a diet containing dried egg white as a protein source. She termed this condition "egg-white injury." Kogl and Tonis isolated the factor from egg yolk and called it biotin. In 1940, Gyorgy and duVigneaud showed that biotin was the same as vitamin H and they determined its chemical structure.

Folic Acid   Wills observed that an autolyzed yeast extract relieved macrocytic anemia in Indian women. A similar anemia in rats was prevented by yeast or liver extract. Day, in Arkansas, called an anemia-prevention factor vitamin M. The main events are listed in Table 8.

Folic acid and vitamin B-12 are both concerned with the prevention of anemia in humans. These vitamins are important for normal growth in animals and birds.

Vitamin B-12 (Cyanocobalamine)   McCollum (1957) reviewed the historical information about anemia. Iron had been used for centuries to cure anemia. Thomas Sydenham (1624–1689) introduced "steel tonic" (iron filings in wine) as a cure for anemia. Clinical experience revealed that certain cases did not respond to iron therapy. Thomas Addison (1793–1860) described "pernicious anemia," for which no remedy could be found. A series of studies starting in 1927 by Whipple and associates to measure the hemoglobin capacity of different foods showed that liver,

Table 8
Discovery of Folic Acid

Date	Persons and Events
1938	Stokstad at California; factor U for growth of chicks; Norris *et al.* at Cornell; factor R, growth of chicks
1939	Hogan and Parrott at Missouri; Vitamin B_c prevented macrocytic anemia in chicks
1940	Snell at Wisconsin—*L. casei* factor needed for growth of bacteria; Mitchell, Snell, and Williams—also needed for *S. faecalis;* called folic acid
1943	Pffeffner, Park Davis, and Stokstad; Lederle crystallized folic acid
1946	Lederle group synthesized folic acid and several other active forms

kidney, and gizzard were most effective. Because people had great difficulty taking the large amount of liver needed to control pernicious anemia, Cohn *et al.* in 1927 prepared a concentrate from liver high in the erythrocyte maturation factor. Later developments leading to the discovery of vitamin B-12, including the extrisic factor of W. B. Castle (1928), which was active in controlling pernicious anemia, are shown in Table 9. Cobalt, the central ion in vitamin B-12, was shown to be a dietary essential for ruminants in 1935 by Underwood and others in Australia. The demonstration in 1928 by Hart *et al.* that copper was necessary for iron utilization in hemoglobin formation illustrates the interrelationships of these nutrients in metabolism.

Other Factors   Other materials for which vitamin activity was claimed were inositol, p-aminobenzoic acid, and certain polyphenols. None of

Table 9
Discovery of Vitamin B-12

Date	Persons and Events
1926	Castle; raw liver gave remission; of pernicious anemia in man
1947	Mary Shorb; liver factor necessary for growth of *L. leichmannii*
1948	E. L. Smith in England, Glasco lab, and Rickes *et al.;* Merck lab isolated a cobalt-containing compound in liver
1951	S. E. Smith *et al.;* injected vitamin B-12, prevented cobalt deficiency in sheep; injection of B-12 brings remission of pernicious anemia and nervous signs in humans; a growth factor for animals; bacterial synthesis in rumen

these compounds have been clearly shown to be dietary essentials for higher animals fed an otherwise adequate diet. While it is possible that other vitamins exist, none have been discovered since 1948.

Nonnutrient Growth-Promoting Substances

Various drugs and hormones, many chemicals, and some fermentation products have been available as feed additives to improve growth or survival of animals. Some of these have proved useful and remain in use while others have not. These include antibiotics, hormones, regulators of metabolism or rumen activity, antimycotics, antiprotozoals, antihelminthics, and pesticides.

Soon after antibiotics were discovered, Moore *et al.* reported in 1946 that streptomycin increased the growth rate of chicks. In 1949, Stokstad and Jukes showed that Aureomycin (chlortetracycline) increased the growth of chicks. The same year, Cunha *et al.* showed that pigs responded to this antibiotic. In 1950, Colby *et al.* found that lambs, and Loosli and Wallace reported that calves, responded favorably to Aureomycin. McGinnis *et al.* observed responses of turkey poults in 1952. Many researchers soon confirmed these findings. Favorable results were noted with most animal species studied except ducks and guinea pigs (National Research Council, 1965).

As more antibiotics and chemotherapeutic agents were developed and used, the federal Food and Drug Administration began to regulate the use of these substances (Siegmund, 1979). By the 1980s, several materials were approved for addition to feeds. Antibiotics approved included bacitracin, bambermycins, carbadox, chlortetracycline, lincomycin, oleandomycin, oxytetracycline, penicillin, streptomycin, tylosin, virginiamycin, and ionophores. Brody *et al.* found that 125 to 250 ppm of copper added to pig feeds would increase the rate of gain; this is used in Great Britain and some other countries but it has not been approved in the United States (Krider *et al.*, 1982).

In recent decades, many drugs have been developed to control internal and external parasites of animals as an aid to maintaining high levels of productivity.

Several hormones, including estrogens, androgens, and thyroid-active materials, have been used to regulate rates of growth, fattening, and milk production. Chemical agents to modify rumen metabolism to reduce methane production or change the proportions of fatty acids produced from cellulose fermentation have been tested. Maynard *et al.* (1979) have described special enzymes, yeast, and bacterial cultures that have been promoted to improve efficiency of production. New advances are continually being made.

Feeding Standards and Feed Values

The first recorded attempt to estimate the comparative usefulness of different feeds for domestic animals apparently was in Bohemia in 1725, where straw feed units were used to compare the value of hays (Tyler, 1975). In his books published in 1808 and later, Thaer quoted some "hay values" and credited his associate Einhoff with having isolated fiber from roughages. Following the development of methods of analyses by Liebig in the 1830s, Grouven in 1859 used feed analyses to formulate feeding standards. Modern-type feeding standards were first developed by E. Wolff in 1864 based on digestible nutrients from the results of feeding trials. Atwater brought the Wolff standards to the attention of American researchers in 1874 and Armsby published them in his book *Manual of Cattle Feeding* in 1880. These were modified by C. Lehman in 1897 and were published by Henry in the first edition of his book *Feeds and Feeding* in 1898. These standards gained wide use in America. In 1914, Haecker pointed out the need to consider the composition of the milk in estimating the requirements of lactating cattle. The standards were modified in the 1915 edition of *Feeds and Feeding* by Henry and Morrison to express requirements of animals and the composition of feeds in terms of digestible protein and total digestible nutrients (TDN), a combined value for the useful energy from carbohydrates, fats, and protein. This standard was used almost exclusively in the United States and some other countries until the 1950s. From extensive metabolism studies with cattle, Kellner compared the value of different feeds to pure starch for fattening animals. These starch equivalents (SE) or values, became the commonly used data in Europe. A feed unit system based on the net energy value of barley was developed by Hansson in Sweden. Several other standards were proposed but did not gain wide use.

From the 1920s to the 1940s, considerable information was published on the mineral and vitamin requirements of various animal species. None of these data were considered in standards of that period. In 1942, the Committee on Animal Nutrition of the National Research Council, under the direction of L. A. Maynard, organized subcommittees to prepare nutrient requirement tables for farm and laboratory animals. The first of these was published in 1944. These reports were revised and updated as new information became available. They are now available for poultry, dairy and beef cattle, sheep, pigs, horses, goats, rabbits, dogs, fish, mink, and foxes, and laboratory animals. The Agricultural Research Council in London published reports on the Nutrient Requirements of Farm Livestock, No. 1, Poultry in 1963, Ruminants and Pigs in 1965. These reports were revised in 1980.

Measuring the Value of Feeds

The earliest pioneers in the analyses of plant materials were Fourcroy and Vauquelin. The most notable successors were Einhoff, Gorham, Hermbsteadt, John, and Vogel, with publications before 1820. In the 1100s, alcohol was collected from a cooled condenser, and in the 1300s, Arnaldo de Villa Nova made alcoholic tinctures of medicines. Fermented drinks were known from early times. Acetic acid was the only acid known to the ancients. In 1703, G. E. Stahl produced glacial acetic acid by freezing vinegar. Spirit of wine (85% alcohol) became available in the twelfth century. Ether was first made from alcohol in 1535 by Valerius Cordus. In the early 1800s, this substance became the solvent for extracting lipids from plant and animal materials. The main events in the development of modern chemical analyses are summarized in Table 10. Justus von Liebig, probably the greatest agricultural chemist of all times, thought that nitrogen content indicated the value of feeds. It was his student Carl von Voit who pointed out to him that the final value could be determined only by animal tests (McCollum, 1957; McCay, 1973).

Table 10
Development of Chemical Analyses of Plant and Animal Tissues

Date	Persons and Events
1760	H. M. Rouelle used various solvents to dissolve parts of plants
1700s	H. Boerhaave and Casper Neumann both used solvents to analyze plants
1811	Gay-Lussac and Thenard analyzed for carbon and nitrogen
1814	Johan F. John extracted plants using alcohol, ether, naptha, and terpentine
1827	Prout made analyses for protein, fats, and carbohydrates
1831	Liebig improved Prout's methods of analyses
1835	J. B. A. Dumas improved combustion analyses methods
1839	Anselm Payne isolated cellulose and assigned the name
1841	F. Varrentrapp and H. Will heated foods in potassium hydroxide to convert the nitrogen to ammonia
1846	E. M. Horsford estimated vegetable fiber
1883	J. Kjeldahl developed the sulfuric acid method for nitrogen analyses
1857	William Henneberg and Frederick Stohmann at the new experiment station at Weende, near Gottingen, devised a system of feed analyses that included moisture, ether extract, nitrogen, ash, and crude fiber, which is still used
1936	Crampton and Maynard proposed the use of cellulose and lignin to replace the variable crude fiber method
1965	P. J. Van Soest devised a detergent extraction system to separate cell wall from digestible plant tissue
1976	Norris *et al.* reported on infrared reflectance spectroscopy to predict composition and quality of forage

Sources: McCollum, 1957; McCay, 1973; Maynard *et al.*, 1979.

In spite of numerous attempts, no one was able to improve or re-place the crude fiber analyses of the Weende system well enough to be accepted until the Van Soest detergent analyses for cell walls was de-vised (Van Soest, 1966). This system for separating digestible from un-digestible parts of plants has now largely replaced the crude fiber-NFE (nitrogen-free extract) system. The search is continuing to find a more rapid, less expensive way to predict the composition and true value of forages for animals. The infrared reflectance spectroscopy analyses promises to be such a method, but it still requires more testing and improvement.

In 1947, Schneider published tables showing the composition and digestibility of feeds for cattle, sheep and goats, and horses and pigs (Schneider, 1947). These tables were supported by a bibliography of the original research up to 1940. Since 1960, the University of Florida has tabulated feed composition data from all Latin American countries.

The Florida data were stored in the computer program of the Foodstuffs Institute under the direction of Lorin E. Harris at the Utah Agricultural Experiment Station, Logan, Utah. This unit also tabulated all of the analyses and digestibility data for the National Research Coun-cil (NRC) U.S. and Canadian feed composition tables and the nutrient requirement reports for various animals published in recent years. Har-ris and Associates organized the International Feedstuffs Institute with headquarters in Utah and cooperating centers in Europe, South Amer-ica, Asia, and Africa. This has great promise for the development of animal nutrition research on a coordinated, worldwide basis. The pro-gram has been reviewed by Harris *et al.* (1980).

Other Advances that Have Aided Animal Nutrition

A number of other elements were important to the development of animal nutrition. These included the organization of various scientific societies and their publication of important discoveries. The main so-cieties that stimulated development in this field are listed in Table 11. The first organizations were in Europe, followed by those in the United States in the nineteenth century. The professional societies were started largely in the early 1900s.

A number of important and classic books, not cited in the refer-ences, contributed greatly to the dissemination of knowledge. Some of these are listed in Table 12.

Table 11
Organizations that Aided Development
of the Sciences

Date Founded	Organization
1645	The Royal Society, Oxford, England
1650	First Scientific Society, Germany
1660	Academie des Sciences, Paris, France
1848	American Association for the Advancement of Science
1863	National Academy of Sciences
1876	American Chemical Society
1905	American Dairy Science Association
1908	American Society of Animal Nutrition
1908	Poultry Science Association, United States and Canada
1912	American Society of Animal Production
1928	American Institute of Nutrition
1916	National Research Council (under the National Academy of Sciences)
1962	American Society of Animal Science

The organization of the land-grant college system associated with the agricultural experiment stations and the Extension Service provided the facilities and opportunities for expansion of the European knowledge and its application to American and Canadian conditions.

The U.S. Land-Grant College System

The land-grant college system was created in 1862 by the Morrill Act of Congress to teach agriculture and engineering to the rural population In 1887, the Hatch Act provided federal support for the creation of the agricultural experiment stations, associated with the land-grant colleges of each state, to develop a body of scientific knowledge for useful college teaching. Federal support for the Agricultural Extension Service in each state was provided in 1914 by the Smith-Lever Act for the purpose of teaching agriculture, home economics, and related subjects to adults on their farms. Land-grant colleges and the U.S. Department of Agriculture were to cooperate in diffusing practical information on agriculture and home economics through instruction and practical demonstrations on farms and in homes to persons not attending land-grant colleges. Over the years, the extension animal, dairy, and poultry science specialists, along with experiment station researchers and land-grant college teachers, have aided in identifying livestock production problems and assisting with their solution.

Table 12
Classic Books that Aided Development of Animal Nutrition[a]

Date	Author	Title
1608	Jean Beguin	*New Light on Chemistry for Beginners,* Paris
1660	Nichasius le Febure	*Traite de chemie,* France
1760	C. Neumann	*The Chemical Works of G. Neumann* (translated by W. Lewis from German)
1777	P. J. Macquer	*A Dictionary of Chemistry*
1804	Thomas Thomson	*System of Chemistry,* Glasgow
1814	J. F. John	*Chemische Tabellen der Pflanzenanalysen,* Nurnberg
1814	Arthur Young	*Annals of Agriculture* (45 volumes)
1809–12	A. D. Thaer	*Principles of Agricultural Crops and Animal Husbandry* (translated)
1834	W. Prout	*Chemistry, Meterology and Function of Digestion* (Bridgewater treatise)
1842	Justus von Liebig	*Thierchemie*
1843	Thomas Thomson	*Animal Chemistry,* Glasgow
1843–44	J. B. Boussingault	*Economie rurale*
1880	H. P. Armsby	*Manual of Cattle Feeding*
1881	Carl von Voit	*Stoffwechsel und Ernahrung*
1895	E. P. Moller	*Cod Liver Oil and Chemistry*
1898	W. A. Henry	*Feeds and Feeding* (1915–1936 with Morrison)
1903	W. H. Jordon	*The Feeding of Animals*
1905	O. Kellner	*Die Ernahrung der Landwirtschafliche Nutztier*
1917	H. P. Armsby	*Nutrition of Farm Animals*
1923	L. B. Mendel	*Nutrition: The Chemistry of Life*
1925	H. P. Armsby and C. R. Moulton	*The Animal as a Converter of Matter and Energy*
1936–56	F. B. Morrison	*Feeds and Feeding*
1945	Samuel Brody	*Bioenergetics and Growth*
1947	B. H. Schneider	*Feeds of the World: Their Digestibility and Composition*
1954	R. W. Swift and C. E. French	*Energy Metabolism and Nutrition* / *Applied Animal Nutrition*
1961	Max Kleiber	*The Fire of Life*
1966	E. J. Underwood	*The Mineral Nutrition of Livestock*

[a]Books not listed in the references cited.

The Federal System for Animal Nutrition Research

The U.S. Department of Agriculture (USDA) was established in 1862. Early research was oriented principally toward plant introductions until 1884, when the Bureau of Animal Industry (BAI) was created. From the bureau has evolved the Agricultural Research Service (ARS), which is currently the principal research arm of the department. The purpose of the ARS is to provide research information for solving food and agricultural problems of broad scope and of high national priority. This service has a major internal research program and also supports, through cooperative agreements, significant amounts of research at the land-grant colleges. In addition, the department supports research via the competitive grants program, which is open to most institutions conducting agriculturally related research. Finally, federal funding is provided to the agricultural experiment stations of the land-grant system through the formula funds that are overseen by the department's Cooperative State Research Service. Additional and significant federal support for agriculturally oriented research is also provided by the National Science Foundation, the National Institutes of Health, and the Department of Energy.

Rumen Metabolism

Hungate's book (1966) cites the discovery of acetic and butyric acid in rumen contents in 1831, but propionic acid was not recognized as an important product of cellulose digestion until Elsden's research in 1945. Methane and carbon dioxide were described by Reiset in 1863 as products of rumen fermentation. Rumen fistulas were first mentioned by Fluorens in 1833, and Colin used them in research around 1850. In 1855, Haubner showed that large amounts of cellulose disappeared as food passed the rumen. Pasteur found in 1863 that bacteria fermented plant material. Zuntz was first (1883) to explain the mechanism of forage utilization by ruminants, and in 1891 he showed that rumen bacteria made preferential use of nonprotein sources. Ruelle found urea in the urine in 1773. Wohler produced urea in 1842 by heating ammonium cyanate. In 1886, Colin found urea in the saliva of cattle, and in 1919, Voltz demonstrated that urea could serve ruminants as a nitrogen source. Following much conflicting research in Germany, Bartlet and Cotton showed that heifers grew faster on a ration in which urea supplied 30% of the nitrogen than on the low-protein basal feed (McCollum, 1957). Rupel, Bohstedt, and Hart, at Wisconsin, found that urea was equal to linseed meal as a supplement to forages and cereal grain for milking cows (1943). It was reported in 1949 that all of the essential amino acids were synthesized in the rumen when urea was the sole nitrogen source (Maynard, 1969). Use of nonprotein nitrogen is now common in ruminant feeds.

Theiler *et al.* expressed the view in 1915 that the vitamin requirements of cattle are so low that adequate amounts may be synthesized by rumen bacteria. Bechdel *et al.* reported in 1926 that cattle did not require a dietary source of vitamin B because of synthesis by rumen bacteria.

Later, studies showed that all eight of the B vitamin complex are synthesized in the rumen. Modern books on animal nutrition review the more recent research on rumen metabolism.

Biographical Sketches

Brief biographical sketches are presented of some of the men who made the greatest contributions to the discoveries of the essential nutrients and early developments in the area of animal nutrition (World Who's Who in Science: From Antiquity to the Present, 1968). Many others made significant contributions to nutrition knowledge, but emphasis is given to those who relate most directly to animal production. The books by Armsby, Brody, Kleiber, Lusk (1933), Maynard (1937), McCay (1973), McCollum (1957), Mitchell, and Morrison (Table 12) give greater detail about the contributions of these and other men.

Henry Prentis Armsby (1853–1921) became director of the Pennsylvania experiment station at State College in 1887 and of the Institute of Animal Nutrition in 1907 (Maynard et al., 1979). He built a respiration calorimeter and did research that led to the net energy system of evaluating feeds and the energy requirements of animals. His books set the standards for teaching and research in animal nutrition in the United States.[1]

Wilbur Olin Atwater (1844–1907), a pupil of Voit, Rubner, and Zuntz, was director of the first agricultural experiment station in the United States—at Wesleyan University Middletown, Connecticut, in 1875 (It was later moved to New Haven). With Gano, he constructed the first human-respiration calorimeter.[2]

Stephen M. Babcock (1843–1931) invented the test for milk fat that bears his name. He made many contributions to dairy chemistry and animal nutrition, including the demonstration in 1905 that sodium is essential for lactating dairy cows.[3]

John Jakob Berzelius (1779–1848) was educated at Uppsala University in chemistry and became professor of botany and pharmacy at Stockholm University in 1807. He discovered several elements, including selenium and silicon, and he invented the system of symbols still used for chemical elements. He determined the atomic and molecular weights of over 1000 chemical substances.

Joseph Black (1728–1799) was professor of chemistry and medicine

1. *J. Nutr.* (1954) **54**, 3.
2. *J. Nutr.* (1962) **78**, 3.
3. *J. Nutr.* (1949) **37**, 3.

in Britain. He served as physician to George III in Scotland. He invented an ice calorimeter, originated the concept of specific heat and latent heat, helped lay the foundation for modern quantitative chemistry, and was the first to show that carbon dioxide (fixed air) could be formed by decomposition as well as by combustion and fermentation. His pupil James Watt invented the steam engine.

Jean Baptiste Boussingault (1803–1887) was a professor of chemistry at Lyons, France, who founded the first experiment station at Bechelbrom in 1836. His books published in 1843 and 1844 dealt with nutrition of cattle, horses, hogs, and other animals as well as with soils, crops, and fertilizers. His research confirmed that animals could not use nitrogen from the atmosphere to supplement inadequate protein in the feed. He was the earliest in animal experimentation in nutrition problems. He demonstrated that common salt was indispensable for the well-being of farm animals and that potassium, calcium, and phosphorus were necessary.[4]

Robert Boyle (1627–1691) was born in Ireland and was educated at Eton and Oxford, where he established a laboratory in 1654. He is best known for his gas laws. He studied chemistry and respiration and postulated that life was a slow-burning process. Much of his extensive writings bear interest for nutritionists. He reported that one fraction of air supported combustion while another did not. He prepared ammonia, gelatin, and SO_2 to preserve legumes. He designed a vacuum pump and a compressed air pump. He developed an autoclave and showed that autoclaved foods did not spoil if air was excluded. He pointed out that body composition could be modified by food and that fish flavor passed into pork. Freezing was recommended for meat preservation and also for preservation of biological specimens.

Georg Brandt (1694–1768) was a Swedish chemist who also had a medical degree. He discovered cobalt in 1735 and that calamine contained zinc. He did research on arsenic and distinguished between potash and soda.

Samuel Brody (1890–1956) was born in Lithuania. He joined the dairy husbandry department at the University of Missouri in 1920. There he made many studies on growth, energy metabolism, and environmental physiology, which were summarized in his book, *Bioenergetics and Growth*, in 1945.[5]

Gustave von Bunge (1844–1920) was born in Dorpat, Germany, and trained at that university as well as in Leipzig and Strassburg in chemistry and medicine. He taught at those universities and moved to Basel

4. *J. Nutr.* (1964) **84**, 3.
5. *J. Nutr.* (1960) **70**, 3.

as Professor of Physiology in 1885. His main research was on iron metabolism.[6]

Henry Cavendish (1731–1810) was an English physician and chemist. He discovered the properties of hydrogen in 1766 and the composition of water and he determined the composition of the atmosphere in 1781. He deduced the composition of nitric acid and synthesized it.

Michel Eugene Chevreul (1787–1889) contributed more about the chemistry of fats than any other chemist. He recognized that fats were an ester of fatty acids and glycerine. His isolation of palmitic acid in 1811 was the first of many fatty acid studies including oleic, stearic, butyric, and caproic, and he studied metallic salts of many of them. He lectured in Paris when he was more than 100 years old.[7]

Charles J. E. Chossat (1796–1848) was a physician in Geneva. He was the first to show that in pigeons fed a diet of wheat and water their bones became deficient in ash but remained normal when calcium carbonate was added.

Earle Wilcox Crampton (1895–1983) was born in Connecticut and obtained his B.S. degree there. Later, he received his M.S. at Iowa State University and a Ph.D. at Cornell University. During his 51-year career, he taught many students basic and applied nutrition at MacDonald College, Canada. His research involved the determination of the nutritional requirements of animals with emphasis on pigs. He studied the utilization of various feeds and their effects on carcass quality. He served on many livestock and nutrition committees in Canada and as a member of the Committee on Animal Nutrition and its subcommittees of the U.S. National Research Council for three decades. He promoted the change from the use of TDN to DE, ME, and NE. With Lorin E. Harris, he prepared the *Atlas of U.S.–Canadian Feeds* and promoted the establishment of the International Feedstuffs Institute.

Adair Crawford (1748–1795), a British physician and professor of chemistry, was the first to recognize strontium. He did research on animal heat, specific and latent heats, and the chemistry of respiration. In 1779, he published a book entitled *Experiments and Observations on Animal Heat and the Inflammation of Combustible Bodies.*

Baron Alex Fredrik Cronstedt (1722–1765) was a Swedish mineralogist. He discovered nickel in 1751 and later zeolite. His classification of the minerals was based on chemical composition.

Conrad A. Elvehjem (1901–1962) was trained at the University of Wisconsin. He was a superb teacher and researcher who trained many graduate students. He discovered that nicotinic acid would cure or pre-

6. *J. Nutr.* (1953) **49,** 3.
7. *J. Nutr.* (1960) **72,** 3.

vent black tongue in dogs. With Hart he discovered that copper is essential for hemoglobin formation. With his students he studied the metabolism of many of the B vitamins.[8]

Erasistrato (of Chios) (c. 304–250 B.C.) perhaps was a grandson of Aristotle. He studied in Athens and founded a school of anatomy in Alexandria. He made many studies on the growth of the body and on digestion.[9]

Ernest Browning Forbes (1876–1966) was trained at Illinois and Missouri. He worked at Ohio before succeeding Armsby as Director of the Institute of Animal Nutrition at Pennsylvania in 1922. His research included mineral balances, energy utilization for growth, lactation and reproduction, and the effects of levels of energy, fat, and protein on the net energy value of feeds.[10]

George Fordyce (1736–1802) was a British physician. He wrote *Elements of Agriculture and Vegetation* (eight volumes) in 1765 and *Treatise on Digestion of Food* in 1791.

Antoine F. F. Fourcroy (1755–1809) was trained in medicine and believed that chemistry was important to medicine. He held the chair in chemistry at the Jardin du Roi. Among other things, he analyzed milk, spring water, blood, and urine. His 10-volume work on chemistry published in 1800 brought together the chemical knowledge of his time.

Casimir Funk (1884–1967) was born in Warsaw, Poland. He joined the Lister Institute in England, where he published a list of deficiency diseases and presented the theory that beriberi, scurvy, pellegra, and possibly rickets were caused by the lack of special substances in the diet that are organic bases he called "vitamines." From his extracts, he isolated nicotinic acid in 1914 but didn't recognize its nutritional significance.[11]

John Gottlieb Gahn (1745–1818) was a Swedish chemist and mineralogist. He trained Berzelius in quantitative analyses. With Scheele, he discovered phosphoric acid in bones in 1770. He discovered selenium in the flue dust of one of his sulfuric acid plants.

Joseph G. Gay-Lussac (1778–1850) was born in Limousin, France, and was trained in Paris. He served as assistant to Fourcroy and Verthollet. He was professor of chemistry at the Ecole Polytechnique in 1809 and at the Jardin des Plantes in 1832. He isolated boron and invented a hydrometer, portable barometer, and the maximum-minimum thermometer. He used litmus paper in studies of acids and bases and used

8. *J. Nutr.* (1971) **101,** 569.
9. "World Who's Who in Science: From Antiquity to the Present," Marquis Who's Who, Chicago, 1968.
10. *J. Nutr.* (1970) **100,** 1013.
11. *J. Nutr.* (1972) **102,** 1105.

the term "neutrality" in the modern meaning. He studied gases with Voit and the fulminates with Liebig and Humboldt. He is considered the father of volumetric analysis.

Edwin B. Hart (1874–1953) was born in Ohio and obtained his first degree at the University of Michigan. He was a chemist at the New York experiment station before becoming head of Agricultural Chemistry at the University of Wisconsin in 1906. He made many contributions to nutrition, including the discovery that copper is essential for hemoglobin formation. He was an especially distinguished teacher.[12]

William Arnon Henry (1850–1932) graduated from Cornell University and became Dean of Agriculture at the University of Wisconsin. His book with Morrison, *Feeds and Feeding*, was the leading textbook on nutrition, feeding, and management of livestock for decades.

Alex Holst (1860–1931) was trained in medicine and became professor of hygiene and bacteriology at the University of Christiana in Oslo in 1892. He was well known for his studies of ship beriberi. With Theodor Frolich (1870–1947), he produced scurvy in guinea pigs and made many studies on infant scurvy (Barlow's disease). Their studies of the antiscorbutic value of food was widely accepted in the medical profession.

Frederick Gowland Hopkins (1861–1947) was a biochemist at Cambridge University. He discovered glutatione in 1921 and showed that histidine is an essential amino acid. Jointly with Eijkman he was awarded the Nobel Prize in medicine for research on vitamins, especially thiamin.[13]

Oscar Kellner (1851–1911) served at Proskau, Hohenheim, and at the University of Tokyo in agricultural chemistry before going to Mockern, Germany, in 1893. He was best known for his respiration studies, which were the basis for his feeding standards for farm animals published in 1905.[14]

Max Kleiber (1893–1976) was born in Switzerland and trained in agricultural chemistry. In 1929, he moved to the University of California at Davis where he built a respiration apparatus. He developed the use of body weight to the 0.75 power instead of surface area to compare energy metabolism of various animals. His research included isotope studies and nutrient utilization in milk secretion.

Antoine-Laurent Lavoisier (1743–1794) has been called the father of nutrition. He discovered that combustion was an oxidation and that respiration supplied oxygen to the tissues for the breakdown of food to produce carbon dioxide and water and release energy for the body. He

12. *J. Nutr.* (1953) **51**, 3.
13. *J. Nutr.* (1950) **40**, 3.
14. *J. Nutr.* (1952) **47**, 3.

built a calorimeter, with Laplace, and showed that respiration is the source of body heat and productive energy. He used sparrows, guinea pigs, and humans in his studies.

John B. Lawes (1814–1900) and Joseph H. Gilbert (1817–1901) were pioneer researchers at the Rothamsted experiment station in England, which is still well known for research in plant sciences. Their studies on the composition of animals used as human food opened an area that is still active today. They showed that animals produce body fat from carbohydrates.[15]

Leonardo da Vinci (1452–1519), an Italian artist and scientist, was educated in Florence. He sparked the Italian renaissance and has been called the father of modern science. In relation to nutrition, he presented these ideas:

> The body or anything that takes nourishment constantly dies and is constantly reborn.
> Where there is life, there is heat.
> If you do not restore day-by-day the nourishment departed, life will fail.
> An experiment is the repetition of a natural process designed to discover the laws of relations presented by science.
> An experiment is never fallacious, only our interpretation of it may be wrong.
> Poor is the pupil who does not surpass his master.

Da Vinci taught Columbus and Amerigo Vespucci and enabled them to sail to America. He was one of the greatest, most versatile geniuses of all times.[16]

Justus von Liebig (1803–1873) was born in Darmstadt. After completing his schooling in Germany, he studied in France under Gay-Lussac. In 1824, he became professor at Giessen and in 1852 at Munich. He organized the first chemical laboratory for students in Continental Europe. He became the foremost organic chemist of his time and was a primary founder of agricultural chemistry. Using his modern methods of organic analysis, he began accumulating knowledge regarding the composition of foods, tissues, feces, and urine. His contention that protein, energy foods, and certain minerals were all that an animal needed for nourishment may explain why the vitamins were not discovered in Germany.

Clive Maine McCay (1898–1967) was born in Indiana and trained at

15. *J. Nutr.* (1966) **90**, 3.
16. "World Who's Who in Science: From Antiquity to the Present," Marquis Who's Who, Chicago, 1968.

the University of Illinois, Iowa State College, and the University of California under C. L. A. Schmidt. After studying at Yale University as an NRC Fellow under L. B. Mendel, he joined L. A. Maynard at Cornell University. His nutrition research involved many species, from insects to fish, laboratory animals, dogs, and farm animals. He was the first to show that the life span of rats could be about doubled by severely restricting the energy intake on an otherwise adequate diet. He is best remembered for his research on utilization of calcium and fluorine, on effects to improve the nutritive quality of bakery products, on the adequacy of U.S. Navy diets, and on the elderly in institutions. He was a stimulating teacher and widely invited to speak about nutrition. His book *Notes on the History of Nutrition* is a classic.[17]

Elmer Verner McCollum (1879–1967) earned his B.A. and M.S. degrees at Kansas University and his Ph.D. at Yale University in organic chemistry. He worked a year with Osborne and Mendel, then accepted a position with E. B. Hart at the Wisconsin experiment station. His research with rats led to the discovery of vitamin A in 1913. In 1917, he accepted a position as head of the Department of Chemical Hygiene and Public Health at Johns Hopkins University. There, with his colleagues, vitamin D was separated from vitamin A. In later research, he described deficiency symptoms in rats for magnesium and manganese in 1931. His textbook, *Newer Knowledge of Nutrition*, first published in 1918, appeared in five editions through 1935. His *History of Nutrition* was also a classic. E. B. Hart said, "No man in our time has contributed as much to the popularization of the subject of nutrition as Dr. McCollum."[18]

François M. Magendie (1783–1855) was a leading French physiologist who taught Claude Bernard. In 1816, he showed that special organic forms of nitrogen were necessary in the diet and that atmospheric nitrogen could not be used. The dogs he fed diets of olive oil, sugar, and water developed ulceration of the cornea, typical of vitamin A deficiency. He recognized an analogous condition in man as a result of a restricted diet. He was the first to use rodents in nutrition research. He described loss of fur and emaciation similar to vitamin B complex deficiency. Magendie was, in fact, the father of the vitamin theory, unrecognized until a century later.[19]

Leonard Amby Maynard (1887–1972) was an undergraduate at Wesleyan University. He received a Ph.D. at Cornell University in chemistry. Starting in 1915, he served as researcher, teacher, and administrator in animal husbandry and in human nutrition and biochemistry. His book, *Animal Nutrition*, first published in 1937, has been widely

17. *J. Nutr.* (1973) **103**, 1.
18. *J. Nutr.* (1970) **100**, 483.
19. *J. Nutr.* (1951) **43**, 3; *Science* (1930) **71**, 315; *Science* (1931) **74**, 456.

used. His research was concerned with the requirements and metabolism of protein, lipids, and minerals, and vitamins with domestic and laboratory animals. As chairman of the NRC Committee on Animal Nutrition, he organized committees in 1942 to develop the nutrient requirement reports for various animals.[20]

Harold Hansen Mitchell (1886–1966) was trained in chemistry. He served at the University of Illinois as teacher and researcher in animal nutrition from 1913 until he retired in 1956. With Tom Sherman Hamilton (1894–1972), he made many contributions in the fields of energy, protein, minerals and vitamin requirements, and metabolism. His two-volume book, *Comparative Nutrition of Man and Domestic Animals*, published in 1963, gives an excellent summary of nutrition up to about 1960.[21]

Baron Frank Morrison (1887–1958) was trained at the University of Wisconsin, where he became professor of animal husbandry. In 1927, he moved to Cornell University and was head of animal husbandry until he retired in 1955. He joined W. A. Henry to author *Feeds and Feeding* and later became sole author. This text had great influence on animal feeding for over half a century.

Thomas Burr Osborne (1859–1929) and Lafayette Benedict Mendel (1872–1935) were codiscoverers of vitamin A with McCollum and Davis. They made extensive studies of seed and leaf proteins using chemical analyses and rat assays. They developed purified diets to study the amino acids lysine, zein, and gliadin with rats. They were the first to grow chicks to maturity in cages. Their studies showing that natural foods contained many nutritionally indispensable factors not detected by chemical methods available stimulated research to discover the vitamins. Osborne was trained in chemistry at Yale.[22] Mendel studied under Chittenden at Yale, who had established the first teaching laboratory in physiological chemistry in United States.[23]

Cornelis Adrianus Pekelharing (1848–1922), professor of hygiene at the University of Utrecht, found that when mice were fed a baked diet of ingredients thought to be adequate, they soon lost their appetite and died within a month. The addition of fresh milk or whey kept the other animals in good health. He pointed out that an unknown substance in very small amounts was essential for life. This 1905 report led to the search for the unknowns. He thought beriberi was caused by a bacterial infection or toxin.[24]

20. *J. Nutr.* (1974) **103**, 3.
21. *J. Nutr.* (1968) **96**, 3.
22. *J. Nutr.* (1956) **59**, 3.
23. *J. Nutr.* (1956) **60**, 3.
24. *J. Nutr.* (1964) **83**, 3.

Joseph Priestly (1733–1804) was an English chemist who also trained as a pastor. He was the author of many works. He invented a method of making soda water in 1771, discovered oxygen in 1774, and knew that green plants produced oxygen in sunlight. He isolated and described nitrous oxide, nitrogen peroxide, ammonia, sulfur dioxide, hydrogen sulfide, carbon monoxide, and hydrochloric acid gas.

William Prout (1785–1850) was an English physician. In studying the relations of physiology to chemistry, he discovered hydrochloric acid in the gastric juice in 1823, that the snake excreted its waste nitrogen as uric acid, and that the developing chick embryo uses the phosphorus from the yolk to form its bones and takes the needed calcium from the shell. He prepared pure urea and determined its composition in 1818. In 1827, he was the first to state that the objective of food analysis was to determine the saccharine (carbohydrate), oily (fats), and albuminous (protein) principles.[25]

William Cummings Rose (1887–1985) received his Ph.D. degree at Yale University under Mendel. He was professor at the University of Pennsylvania Medical School, the University of Texas, Galveston, and the University of Illinois. He discovered threonine and defined its structure. This last essential amino acid to be found made it possible to formulate purified diets with only pure amino acids in place of proteins. He contributed greatly to the knowledge of amino acid requirements of rats and humans and to the metabolism of various nitrogen compounds.[26]

Max Rubner (1854–1932), the most illustrious pupil of Voit, developed a number of modern concepts relating to energy metabolism. He established that carbohydrates and fat were interchangeable in metabolism based on energy equivalents. This is the isodynamic law of Rubner. His refined techniques enabled him to establish the relation of surface area of an animal and its resting heat of metabolism (the surface area law). He observed and defined the specific dynamic effect (or action) of foods. These concepts corrected some of Liebig's errors and set the stage for modern research on energy metabolism in Europe and the United States. Rubner served as professor of physiology at the University of Berlin.

Santorius (Santorio) (1561–1636), an Italian physician and professor at Padua, would eat while balanced on a steelyard to measure "insensible perspiration," which he did for some years. Galileo, a friend, had invented an air thermometer, which he used to record body temperature. His example led to studies on energy metabolism.

25. *J. Nutr.* (1977) **107,** 15.
26. *J. Nutr.* (1981) **111,** 1311.

Karl Wilhelm Scheele (1742–1786) was trained in chemistry to become a pharmacist. In 1771, he learned that the minerals in bones consisted of calcium phosphate. He isolated uric acid in 1776 and was the first to isolate hydrochloric acid from the stomach, showing that chlorine was essential. He discovered glycerine, tartaric, oxalic, citric, and malic acids. He isolated the toxic gases, hydrogen cyanide, and hydrogen sulfide and discovered copper arsenite (Scheele's green). In 1784, he discovered that sulfur was a constituent of the protein of animal tissues.

Raymond W. Swift (1895–1975) was trained at the Massachusetts Agricultural College, Penn State University, and the University of Rochester. He joined E. B. Forbes in 1922 and did research on energy metabolism, promoting the use of digestible energy to replace TDN, and he contributed greatly to the evaluation of foods for animals and humans.

Albert Daniel Thaer (1752–1828) was a physician to George III, King of England and of Hanover. In 1802, he converted his estate in Hanover to the first agricultural institute. In 1804, he became head of a new state agricultural institute at Moglin. Heinrich Einhof, a chemist, worked with him mostly on analyses of farm crops. Thaer's books, published between 1809 and 1812, included tables of "hay equivalents," an early attempt to evaluate different animal feeds.

Louis Nicolas Vauquelin (1763–1829) was a French chemist, a student of Fourcroy. He discovered the first amino acid, asparagine. Among his extensive writings were his *Dictionaire de Chimie et de Metalurgie*, published in 1815 in six volumes, and over 60 memoirs with Fourcroy.

Carl von Voit (1831–1908) has been called the father of metabolism research. He studied under Liebig at Munich and also was a student of Josef von Pettenkofer. He became the leading investigator of metabolism of his time, conducting many studies with dogs and men. Being well trained in biology, he pointed out that the chemical methods developed by Liebig could suggest the probable nutritive worth of a substance, but to be sure they had to be tested with animals or man. He first pointed out that the level of metabolism determined the oxygen consumption. In 1881, he stated that it would be best if pure chemical compounds could be fed: protein, fat, sugar, starch, and ash constituents. His students included Max Rubner, W. O. Atwater, and Graham Lusk.

Nathan Zuntz (1847–1920) was trained as a physician and became a teacher and researcher at Bonn and later at Berlin. He developed the first portable respiration apparatus and did much work on physiological problems with farm animals.[27]

27. *J. Nutr.* (1955) **57**, 3.

References

Blaxter, K. L. (1962). "The Energy Metabolism of Ruminants." Thomas, Springfield, Illinois.

Brody, S. (1945). "Bioenergetics and Growth." Reinhold, New York.

Darby, W. J., Ghaliowngui, P., and Grivetti, L. (1977). "Food: The Gift of Osiris." Vols. 1, 2. Academic Press, New York.

Eddie, W. H. (1941). "What Are the Vitamins?" Reinhold, New York.

Harris, L. E., Jager, F., Leche, T. F., Mayr, H., Neese, U., and Kearl, L. C. (1980). International Network of Feed Information Centers, Publ. 5.

Hungate, R. E. (1966). "The Rumen and Its Microbes." Academic Press, New York.

Jukes, T. H. (1980). Vitamin K: A reminiscence. *Trends in Biol. Sci.* **3**, 140.

Krider, J. L., Conrad, J. H., and Carroll, W. E. (1982). "Swine Production." McGraw-Hill, New York.

Lusk, G. (1932). "The Elements of the Science of Nutrition." W. B. Saunders, Philadelphia.

Lusk, G. (1933). "Nutrition." Paul B. Hoeber, New York.

McCay, C. M. (1973). "Notes on the History of Nutrition." Hans Huber, Bern, Switzerland.

McCollum, E. V. (1933). "A History of Nutrition." Houghton Mifflin, Boston.

Maynard, L. A. (1937). "Animal Nutrition." McGraw-Hill, New York.

Maynard, L. A., and Loosli, J. K. (1969). "Animal Nutrition." McGraw-Hill, New York.

Maynard, L. A., Loosli, J. K., Hintz, H. F., and Warner, R. G. (1979). "Animal Nutrition." McGraw-Hill, New York.

National Research Council (1965). First International Conference on Antibiotics in Agriculture, National Academy of Sciences, Publ. 397, Washington, D.C.

Schneider, B. H. (1947). "Feeds of the World: Their Digestibility and Composition." West Virginia Agricultural Experiment Station.

Siegmund, O. H., ed. (1979). "The Merck Veterinary Manual," 5th ed. Merck and Co., Rahway, New Jersey.

Tyler, C. (1975). Albrecht Thaer's hay equivalents: Fact or fiction? *Nutr. Abstr. and Revs.* **45**, 1.

Van Soest, P. J. (1966). Nonnutritive residues: A system of analysis for replacement of crude fiber. *A. J. Ass. Official Anal. Chem.* **49**, 546.

Vickery, H. B., and Schmidt, C. L. A. (1931). The history of the discovery of the amino acids. *Chem. Revs.* **9**, 169.

Weast, Robert C., ed. "CRC Handbook of Chemistry and Physics." CRC Press, Cleveland, Ohio.

Williams, H. H. (1978). "A History of the American Institute of Nutrition."

"World Who's Who in Science: From Antiquity to the Present." Marquis Who's Who, Chicago, 1968.

A Glossary of Terms Commonly Used in the Animal Industries

Timothy H. Blosser

Agricultural Research Service
U.S. Department of Agriculture
Beltsville, Maryland
(Retired)

Introduction

Each profession, industry, and vocation has terms that are unique to it. Animal producers use words and phrases that are understandable and

Handbook of Animal Science   Copyright © 1991 by Academic Press, Inc.
All rights of reproduction in any form reserved.

tain terms are unique to the husbandry of specific species. This chapter presents definitions and explanations that describe the meaning of the words and phrases widely used by those involved in some phase of animal production, animal product processing, or in servicing the animal industries. Emphasis is placed on terms used in the production of food-producing mammals. Since subject matter covering various aspects of animal production is presented in other chapters of this text, meaningful to others in the same or similar fields of interest but, in some cases, not intelligible to those in other professions and vocations. Certain terms common to those subject matter areas are not covered in this chapter.

A

Abomasum   The fourth or true stomach of the cow, comparable in function to that of simple-stomached animals.

Abortion   Expulsion of the fetus (or fetuses) by a pregnant female before the normal end of a pregnancy.

Acceptability   The extent to which an animal will consume a feed placed before it. Term also used in indicate degree to which animal products are in demand by consumers.

Accredited areas   A township, county, state, or states where bovine tuberculosis has been reduced to less than 0.5% of its cattle population according to the USDA official test.

Accredited herd   A herd of cattle that has been tested for tuberculosis and/or brucellosis and found free of the disease on two successive annual tests or three semiannual tests. Tests must be in accordance with rules and regulations of the Animal and Plant Health Inspection Service, USDA.

Acid detergent fiber (ADF)   Fiber (principally lignin and cellulose) extracted from a feedstuff using a chemical technique employing an acidic detergent. Used to evaluate the quality of a forage.

Acidosis   A condition of the blood which is brought about by an increase in acidity as a result of metabolic processes. There is a consequent reduction in the amount of alkali in the blood.

Adipose tissue   Fatty tissue of animals.

Ad libitum (ad lib)   Term used to reflect the availability of a feed on a free-choice basis.

Aerobic   A biological process requiring the presence of oxygen.

A-frame   Hut, shaped like an A. Used by a sow and her litter on pasture.

Afterbirth   The fetal membranes that attach the fetus to the membranes of the pregnant female and which are normally expelled from the female within 3 to 6 hr after parturition.

Agalactia   Failure to secrete milk following parturition.

Aged cow   Show ring classification of a dairy animal that is 5 or more years old. Also called a mature cow.

Aged herd (swine shows)   Consists of a boar and three sows eligible to show in the junior yearling, the senior yearling, or the aged class.

Agricultural Research Service (ARS)   The principal research agency of the USDA. The agency has scientists at 122 locations in the United States.

Alimentary canal   The tubulous passageway leading from the north to the anus. It includes the mouth, pharynx, esophagus, stomach, small intestines, large intestines, and rectum.

All-American cattle   The most outstanding dairy animals of a specific breed in the various show ring classifications as selected annually by a panel of qualified cattle judges.

Allele   One of a series of two or more alternative forms of a gene occupying a specific point on a chromosome. Alleles influence the same character.

Alternate A.M.–P.M. test   A plan for production testing of dairy cows under the National Cooperative Dairy Herd Improvement Program whereby the milk is sampled by a supervisor at an A.M. milking in one month and a P.M. milking the next month as contrasted with sampling both A.M. and P.M. milkings each month.

Alveolus   A very small structure found in the mammary gland, almost spherical in shape, lined with a single layer of epithelial cells, in which milk is manufactured by the female.

Ambient   That which surrounds (e.g., the temperature around us is the ambient temperature).

American Dairy Science Association   A professional society composed of dairy scientists. The official publication of this society, the *Journal of Dairy Science*, is a means used to publish research findings related to dairy production and processing of dairy products.

American Society of Animal Science   A professional society composed of animal scientists. The official publication of this society, the *Journal of Animal Science*, is a means used to publish research findings related to cattle, swine, sheep, and horse production and the processing of meat and meat products.

American Veterinary Medical Association   A professional society composed of veterinarians. This society publishes two journals, the *Journal of the American Veterinary Medical Association* and the *American Journal of Veterinary Research*.

Amino acid   An organic compound containing nitrogen. Amino acids are the building blocks for proteins, the compounds that are basic to animal tissue structure.

Amnion   The innermost fetal membranes, which form a fluid-filled sac for the protection of the embryo.

Anaphylactic shock (anaphylaxsis)   A state of immediate hypersensitivity following sensitization to a foreign protein or drug.

Anemia   A condition of the blood in which the red corpuscles are reduced in number and/or size or are deficient in hemoglobin.

Animal and Plant Health Inspection Service   The agency of the USDA responsible for administering all federal programs related to animal and plant health.

Anorexia   Lack of appetite.

Antibiotics   Chemical compounds antagonistic to some forms of life produced by and obtained from certain living cells. Antibiotics are used in animal and human medicine to combat certain disease organisms and in animal feeding to enhance performance.

Antibody   A substance produced in an animal as a protective mechanism to combat invasion by a foreign material (antigen).

Antigen   A high-molecular-weight substance (usually protein) that, when foreign to the bloodstream of an animal, stimulates the formation of a specific antibody.

Anus   The posterior opening of the alimentary canal.

Arm   In cattle, that part of the foreleg below the shoulder and above the knee.

Artificial insemination   The injection of mechanically procured semen into the reproductive tract of the female without coition and with the aid of mechanical or surgical instruments.

Artificial vagina   A device simulating the female vagina used for the collection of semen.

As fed   Refers to feed as it is consumed by an animal.

Ash (mineral matter)   The inorganic or mineral matter of a substance that remains after heating the substance to 600° F for 2 hr.

Atrophy   A defect or failure of nutrition or physiological function manifested as a wasting away or diminution in size of cell, tissue, organ, or part.

Average daily gain (ADG)   The average amount of daily live weight increase as applied to farm animals.

B

Baby-beef   Strictly choice or higher grade, fat, young cattle varying in age from 8 to 15 mo and weighing 650–900 lb.

Back   That part of the upper portion of the animal's body between the shoulders and the hip bones.

Backfat   A term used in swine production that refers to the amount of fat deposited along the back in animals of either sex. Usually measured at typical market weights of 200–230 lb. An excessive amount of backfat is considered to be an undesirable trait.

Bacon type hog   A hog with genetic characteristics that permit the production of carcasses yielding a relatively great amount of bacon and a relatively small amount of lard.

Bacterial count   The number of bacteria in a food sample. Can refer either to only viable bacteria or to both viable and nonviable, depending on the technique used.

Bacterial diseases   Infectious diseases caused by bacteria.

Baceteriostatic   A substance that prevents the growth of bacteria but does not kill them.

Bagging a ewe   Looking at or feeling the udder of a pregnant ewe to estimate how close she is to lambing.

Balanced ration   The feed or combination of feeds that will supply the daily nutrient requirements of an animal.

Baled hay   Hay that has been compressed into either rectangular or round packages. Rectangular bales normally weight between 60 and 100 lb. Large round bales weigh as much as 1000 lb. Bales are bound either with wire or twine.

Band   In range terms, refers to a group of ewes numbering 700 to 1200.

Barrel   That part of the animal's body between the fore and rear legs.

Barrow   A male swine animal that was castrated at an early age before reaching sexual maturity and before developing the physical characteristics peculiar to boars.

Basal metabolism   The chemical changes that occur in the cells of an animal in the fasting or resting state when it uses just enough energy to maintain vital cellular activity, respiration, and circulation.

Beef   The meat from cattle other than calves (the meat from calves is called veal).

Beef Improvement Federation   A federation of organizations interested in the genetic improvement of beef cattle. Makes recommendations regarding performance testing procedures.

Behavior   The reaction of an animal to internal and external stimuli, which is an attempt by the animal to adjust or adapt to the situation.

Birth canal   The channel for parturition formed by the cervix, vagina, and vulva.

Blemish   Any defect or injury that mars the appearance of, but does not impair the usefulness of, an animal.

Blind quarter   In cattle, a mammary gland quarter that does not secrete milk or one that has an obstruction in the teat that prevents the removal of milk. A nonfunctional mammary gland.

Bloat   A disorder of ruminants characterized by the accumulation of gas in the rumen.

Blood groups   Blood types that are immunologically different due to hereditary differences in antigens or antigenic factors carried on red blood cells.

Bloom   A term commonly used to describe the beauty and freshness of a cow in early lactation. A dairy cow in bloom has a smooth hair coat and presents evidence of milking ability. The term is also used with other species of farm animals to apply to an animal in excellent condition.

Bluestone   Copper sulfate.

Boar   An uncastrated male swine animal of any age.

Boar odor   The objectionable odor coming from the meat of a mature boar when it is cooked.

Bolus   A masticated morsel of food ready to be swallowed.

Bomb calorimeter   An apparatus used to measure the heat of combustion of foods, feeds, or fuel.

Boston butt   A wholesale cut of pork consisting of the upper portion of the shoulder area.

Bovine   Pertaining to cattle. The genus *Bos*.

Brand   A marking made with a hot iron, branding fluid, or a very cold instrument for identification purposes on the hide (or wool) of animals.

Breast   That front portion of the sheep or cattle body below the neck, above the foreleg, and in front of the shoulder. More often termed the brisket in cattle.

Breech   The buttocks. A breech presentation at birth is one in which the rear portion of the fetus is presented first.

Breed   Animal having a common origin and characteristics that distinguish them from other groups within the same species.

Breed character   A combination of masculinity or femininity with ideal breed type features.

Breeding value   Genetic worth of an animal's genotype for a specific trait.

Brisket   That portion of the sheep or cattle body located just in front of the forelegs. See also *breast*.

British breeds   Those breeds of beef cattle native to Great Britain, such as Hereford, Angus, and Shorthorn.

Broad-spectrum antibiotic   An antibiotic that is active against a large number of microbial species.

Broken mouth   Sheep who have lost some, but not all, of their teeth.

Broken udder   A term used to indicate an udder that is pendulous and loosely attached.

Browse   Woody or bushy plants upon which livestock (and some wild animals) feed.

Buck   Male sheep. Generally refers to those of breeding age.

Bulk tank truck   A vehicle bearing a large, stainless steel tank in which milk from farm bulk milk tanks is transported to a processing plant.

Bull   Male cattle. The term is also used for males of several other species.

Bull index (sire index) A measure of the inheritance of milk and/or milk nutrient production that a bull tends to transmit to his daughters.

Bummer lamb A lamb orphaned for one of a variety of reasons and being raised by hand or by "bumming" milk from other ewes.

Burdizzo pincers A device resembling a pliers that is used to crush the spermatic cord and thus to castrate the male animal in what is called "bloodless castration."

C

Calf starter A dry concentrate feed especially formulated for use with young calves from birth to 3 or 4 mo of age.

California Mastitis Test A cowside test for determining whether or not a cow is afflicted with mastitis. A stream of milk is placed in a small cup containing a chemical reagent. The reaction observed indicates the absence or presence and/or severity of mastitis.

Calving The act of a cow giving birth to a calf.

Calving interval Time between freshenings (parturitions) in cows.

Canner In cattle marketing, an animal of the poorest grade.

Cannibalism Consists of one pig biting the tail of another. Most frequently observed in closely confined, growing-finishing pigs.

Carcass merit The value of a carcass for consumption.

Carotene A fat-soluble pigment widely distributed in nature and synthesized in plants only. The precursor of vitamin A.

Casein A group of proteins similar in structure and constituting the principal proteins in milk. Comprises about 3% of normal cow's milk.

Castration Removal of the testes from a male animal.

Cattle guards Devices consisting of round metal bars placed over a shallow pit and placed at ground level, replacing a gate in roads or lanes and sufficiently strong enough to bear the weight of a car, wagon, or truck passing over them but which deter cattle, hogs, sheep, and most horses.

Cattle squeeze A cratelike mechanical device used to immobilize cattle for the purpose of administering treatment or surgery.

Cellulose The most abundant organic material on earth. A chemically complex carbohydrate making up most of the structural parts of a plant such as the cell walls. Digestive tract enzymes do not digest cellulose.

Central farrowing house A structure (barn) into which pregnant sows are brought to give birth to their young. These buildings normally have special features related to farrowing needs and may be environmentally controlled.

Central testing station Locations to which young swine, beef cattle, or lambs are brought to test their genetic ability to gain weight efficiently when housed and fed under uniform conditions. Animals of both sexes may be tested, and, in some cases, test animals or sibs are slaughtered to determine carcass merit.

Challenge feeding (also called lead feeding)   A feeding system in which an abundance of high energy feed, above what is apparently needed, is provided to cows early in lactation.

Channel Island breeds   Jerseys and Guernseys, dairy breeds developed in the Channel Islands, located in the English Channel between England and France.

Chest floor   That portion of the animal's anatomy on its underside between the front legs.

Chin-ball marker   A device attached to the chin of a surgically altered teaser bull. The device contains ink and is used to identify cows in estrus (heat).

Chine   That portion of the topside of a bovine's anatomy immediately to the rear of the point at which the shoulder blades meet (i.e., the withers) and in front of the loin.

Chromosomes   Structures in the cells of living animals that carry the hereditary factors, or genes. They occur in pairs in body cells and are reduced to one of each pair in reproductive cells.

Chuffy type hog   A type of hog that is relatively small, thick, compactly built, very short of leg and, at the time of marketing, carries a relatively large proportion of its body weight as fat.

Class I milk   Milk used in fluid milk products. Dairymen receive the highest price for this milk.

Class II milk   Milk in excess of Class I needs and used for manufacturing purposes. Dairymen receive a lower price for this milk.

Classified pricing plans   A schedule of prices paid for milk depending on whether it is used for fluid milk or for manufacturing purposes.

Cleaned   A lay term used by cattle producers to indicate that a parturient cow has shed her afterbirth.

Cleaned-in-place (CIP)   A cleaning system for milk pipelines that involves washing them in-place with cleaning solutions.

Clip   The aggregate of all fleeces from a flock.

Closebreeding   An intense form of inbreeding, such as the mating of brothers and sisters or parents and offspring.

Cod   The fatty tissue located at the back and on the underside of a steer in that portion of its anatomy where the testes were located prior to castration.

Colostrum   The first milk produced by the female immediately after giving birth to young. It is highly nutritious and a rich source of antibodies.

Colt   A young male horse of up to about 4 yr of age.

Commercial herds   Herds maintained for the purpose of producing meat, milk, wool, or other products for sale in commercial channels.

Compaction   Closely packed feed in the stomach and intestines of an animal causing constipation and/or digestive disturbances.

Compensatory growth   Increased growth rate in response to previous undernourishment that an animal has recently experienced.

Complete ration   A blend of all feedstuffs (forages and grains) in one feed. A complete ration fits well into mechanized feeding and the formulation of least-cost rations.

Complete self-fed rations   Rations that are well balanced and nutritionally available to animals (most commonly swine) on a cafeteria basis.

Concentrate   An animal feed relatively low in fiber and high in energy, such as a cereal grain or oilseed meal.

Condition   Refers to the amount of flesh (body weight), quality of hair coat, and general health of animals.

Confinement feeding   Refers to feeding in limited quarters, usually under roof and on slotted floors.

Conformation   The body form or physical traits of an animal. Its shape and arrangement of parts.

Controlled estrus   Regulation of the heat period in farm animals by the administration of hormones.

Corn–hog ratio   The number of bushels of corn required to equal in value 100 lb of live hog.

Corral   A fenced enclosure in which cattle are contained.

Cotyledon   An area where the membranes of the fetus and the uterine lining are in close association such that the nutrients can pass to and wastes can pass from the circulation of the developing young.

Cow   A mature, female bovine. The term also applies to female buffalo and to several nondomesticated species.

Cow-calf operation   A beef cattle raising enterprise that specializes in raising beef calves for subsequent growing and fattening.

Cow-hocked   A condition of a cow in which the hocks are close together and the fetlocks (located just above the hoofs) are wide apart.

Cow pool   A business organization or cooperative that cares for and milks cows in a centralized location.

Crampiness   A condition among cattle, especially bulls in confinement, in which involuntary contractions of muscles of the hind legs result in pain and discomfort. Affected animals shift from foot to foot.

Creep   An area accessible to small calves, lambs, or pigs, but not to adult animals, to enable the young to consume additional feed to supplement nutrients obtained from nursing.

Crest   The top of the neck of a male bovine, which, subsequent to puberty, becomes more prominent and, at maturity, attains maximum prominence. An evidence of masculinity in bulls.

Crimp   Natural waviness of the wool fiber.

Critical temperature   The atmospheric temperature below which vasomotor control of the body temperature (e.g., vasoconstriction, vasodilation, sweating, etc.) ceases to maintain body temperature.

Crops   That portion of the animal's anatomy immediately behind the shoulders in the fore rib area.

Crossbred   An animal produced by crossing two or more breeds.

Crossbreeding   Mating systems in which hereditary material from two or more breeds is combined.

Crude fiber   The portion of feedstuffs identified by proximate chemical analysis that is composed of cellulose, hemicellulose, lignin, and other polysaccharides. These compounds make up the structural and protective parts of plants. Crude fiber levels are high in forages and low in grains.

Crude protein   In the proximate analysis of feedstuffs, a figure representing protein content arrived at by multiplying the total nitrogen in a feed by 6.25.

Cryptorchidism   In the male, a condition where the testes are retained in the body rather than descend into the scrotum. Results in infertility.

Cubes   Small packages of hay or other feeds that are used in animal feeding. Typical cubes are 1¼ in square by 2 to 3 in long.

Cud   A portion of feed previously swallowed by the ruminant that is returned to the mouth from the stomach to be chewed a second time.

Culling   The process of eliminating nonproductive or undesirable animals.

Cutability   Fat, lean, and bone composition of meat animals. Used interchangeably with yield grade.

Cutting chute   A narrow chute, used in animal handling and movement, that allows animals to move in single file with gates placed such that animals can be directed into pens along the side of the chute.

D

Dairy character   Physical traits in a dairy cow that suggest high milking ability.

Dairy Herd Improvement Association (DHIA)   An association of dairymen (county, state, national) organized to administer production testing programs.

Dam   The female parent.

Days dry   As used in DHIA records, the number of days dry prior to calving at the start of the production record listed.

Deglutition The act of swallowing.

Dehorning The removal of horns from animals.

Dentition Refers to the dental pattern in various species of animals. A dental formula includes both the numbers and types of teeth, temporary and permanent, in both the upper and lower jaw.

Dewlap In the cow, the flap of loose skin running from the throat to the brisket.

DHIR A type of production testing program in dairy cattle approved by purebred breed associations in which standard DHIA requirements are fulfilled, as well as certain requirements of the breed association.

Diarrhea An abnormally frequent discharge of fecal matter from the bowel. The material discharged is of a very fluid nature. The condition is also called scours.

Digestibility That percentage of food ingested which is absorbed into the body as opposed to that which is excreted as feces.

Digestion The process of converting feeds in an animal's digestive system into simpler compounds that can be absorbed through the wall of the alimentary tract.

Diploid Cells with two members of each chromosome pair. Is characteristic of body cells of all higher animals.

Disqualification A deformity or one or more serious defects that makes an animal unworthy of an award in a show.

Dock In sheep, that portion of the tail setting immediately to the rear of the rump, which is where the sheep's tail originated before its removal by docking.

Docking Removal of the tail. A common practice in lambs.

Dodge Refers to the act of separating various kinds of sheep by means of some mechanical separating device such as a dodge gate.

Doe A sexually mature female goat.

Doeling A female goat from its first to its second birthday or from the first birthday until first parturition.

Dominant One of a gene pair that manifests an effect wholly or partly to the exclusion of the other member of the pair.

Double muscling A hereditary characteristic of certain cattle in which there are bulging muscles of the shoulder and thigh, a very rounded rear end as viewed from the side, a wide but shallow body throughout, appearance of intramuscular grooves, and fine bones. Technically known as muscular hypertrophy.

Drenching A common method of administering medicine to an animal by mixing the medicine with water, elevating the animal's head, and giving the dose from a bottle.

Dress out Removing the offal from a slaughtered meat animal so that the carcass becomes an edible product.

Dressing percentage   Carcass weight/live weight.

Drop band   Ewes that are expected to lamb soon (i.e., within 1 to 2 weeks), also called "heavies."

Dry cow   A cow that is not producing milk.

Drying off   The act of causing a cow to cease lactation in preparation for her next lactation.

Dry lot feeding   Feeding animals in an enclosure away from pasture.

Dry matter   The moisture-free content of feeds.

Dry period   Nonlactating days between lactations.

Dual-purpose cattle   Cattle bred for both milk and beef production in contrast to dairy cattle (bred for milk production) and beef cattle (bred for beef production).

Dummy   Lamb that is slow to react, sluggish, often won't or can't nurse its mother. Also called "crazy lamb" or "daft lamb."

Dung   The feces (manure) of farm animals.

Dwarfism   Undesirable hereditary small size in animals. There are many forms of dwarfism, but the best-known one occurs in beef cattle and is characterized by a short head, short legs, and a "pot belly." It is a recessive, genetic characteristics.

Dystocia   Abnormal or difficult labor at parturition, causing difficulty in delivering the fetus and placenta.

E

Early lambs   Lambs born early enough to go to market as spring lambs.

Ear-notching   A notch or series of notches made in the ear of an animal as a means of identification.

Ear tab or tag   An identification tab or tag clipped to the ear of an animal.

Edema   The presence of abnormally large amounts of fluid in the inter-cellular tissue spaces of the body, as in swelling of mammary glands commonly accompanying parturition in many farm animals.

Ejaculate   Refers to the semen ejected from the male at the time of copulation or to the semen collected as a part of the procedure in artificial insemination.

Elastrator   An instrument for use in stretching a specially made rubber ring, which may be used in dehorning cattle as well as in castrating cattle and sheep.

Electro-ejaculation   Process by which an electrical device is used to obtain semen from a male. The procedure is used to obtain semen from bulls physically incapable of mounting and/or ejaculating with normal stimuli.

Emasculator   An instrument used for docking sheep whereby the tissue at the base of the tail is crushed in the area where the tail is to be cut, thus reducing the loss of blood that occurs when usual procedures are used in docking.

Embryo   A young organism in the early stages of development. In animals, the period from fertilization until life is noted.

Embryo transplant   Artificial transfer of the embryo from the natural mother to a recipient female by mechanical means.

Energy (gross)   The amount of energy manifested as heat when the feed is completely oxidized in a bomb calorimeter.

Energy (metabolizable)   That part of the gross energy which is not carried off in the urine, feces, or gases.

Energy (net)   That part of the metabolizable energy left for growth or product synthesis after subtracting the work of digestion.

Ensilage (silage)   Plant material preserved in a wet state subsequent to fermentation in a silo.

Ensiling   Preparing and storing plant material in a silo.

Ergot   A fungus found in feeds that is toxic to animals.

Eructation   The act of belching or casting up gas from the stomach.

Escutcheon   The area at the rear of a cow that extends upward just above and in back of the udder where the hair turns upward in contrast to the normal downward direction of hair.

Esophageal groove   In ruminants, a semicanal providing passage for fluid materials from the esophagus to the abomasum in young animals. In mature animals the function of the esophageal groove is questionable.

Estrous cycle   The period from one estrus, or heat period, to the next.

Estrus   The period of heat or sexual excitement in the female.

Etiology   The science or study of the causes of disease, both direct and predisposing, and the mode of their operation.

European breeds   In beef cattle, those breeds native to Great Britain or continental Europe, such as Hereford, Angus, Shorthorn, Charolais, Simmental, Limousin, and others.

Eviscerate   The removal of internal organs during the slaughtering process.

Ewe   A sexually mature female sheep. A ewe lamb is a female sheep prior to sexual maturity.

Excrement   Waste material discharged from the body, especially from the alimentary canal.

Exotic   The term used to describe animals foreign to a region.

Experiment station (state agricultural experiment station)   Research organization at each state land-grant university responsible for conducting agricultural research.

Expected progeny differences (EPD)   A term used in beef cattle sire evaluation programs. It is an estimate of how future progeny of a sire are expected to perform relative to progeny of reference sires when both are mated to comparable cows.

Extension Service   An agency in the USDA responsible for national coordination of the activities of the State Cooperative Extension Services in distributing useful information to users (i.e., farmers, etc.).

External parasites   Small insects, such as lice, mites, ticks, flies and mosquitoes, commonly found on the skin of animals.

F

F_1 generation   The first filial generation, or the first-generation progeny, following the parental, or P_1, generation.

F_2 generation   The second filial generation progeny resulting from mating F_1 generation individuals.

Face   That portion of the animal's anatomy below the forehead and above the muzzle and nose.

Facing ("wigging" in Australia)   Trimming the wool around a ewe's face so that she will sense cold, seek shelter, and thus protect the lamb from cold.

False heat   The display of heat (estrus) by the pregnant female.

Farrowing   The act of a sow giving birth to her young.

Farrowing crate   A small enclosure for confining a sow and her litter in such a way that the sow may farrow normally, without hazard to the newborn pigs, and which provides them with space of their own on either side of the sow's compartment.

Farrowing pen   A specially designed pen in which sows are placed at the time of farrowing. These pens contain guard rails to prevent the sow from laying on newly born pigs.

Fasting   Abstaining from food. Certain types of experiments involve withholding food from the test animal for varying periods of time.

Fat-corrected milk (FCM)   A means of evaluating milk production records of different dairy animals and breeds with widely varying levels of production and milk consumption on a common basis energywise.

Fat lamb   A lamb ready for slaughter market.

Fecundity   Efficiency of an individual in production of young. Animals that bring forth young frequently, regularly, and, in the case of those that bear more than one offspring at a birth, in large numbers, are said to be fecund.

Federal Milk Marketing Order   A milk marketing regulation issued by the U.S. Secretary of Agriculture. It establishes minimum prices paid by processors to producers for milk.

Federal Order Administrator   An appointee of the U.S. Secretary of Agriculture who supervises a Federal Milk Marketing Order.

Federal Quarantine Center   A federal (USDA) animal importation facility located at Fleming Key, near Key West, Florida, where imported animals are quarantined to make sure that they are disease-free before they are moved to the farm location of the person who purchased them.

Federal Register   An official document of the federal government in which recommendations and proposals of various federal agencies are published for consideration by the public.

Feed bunk   A structure, usually wooden or metal, in which feed is placed for subsequent consumption by animals.

Feed efficiency   The units of feed consumed per unit of weight increase or unit of production (milk, meat, eggs).

Feeder cattle   Young cattle fed for economical growth and ultimately sold to cattle feeders for fattening and finishing in the feedlot.

Feeder hogs   A market class of hogs weighing between 120 and 180 lb and intended for subsequent fattening.

Feeder lambs   Young animals under 1 yr of age that carry insufficient finish for slaughter purposes but that show indications of making good gains if placed on feed.

Feeding stalls   Stalls in which test animals can be fed individually so that efficiency of gain or production can be measured. Sometimes stall-fed animals gain more uniformly than is the case with group-fed animals.

Feeding standards   Charts and tables showing the recommended daily nutrient allowances for farm animals for various functions, such as growth, maintenance, lactation, etc.

Fell   A thin membrane found between the skin and carcass. Upon exposure to air, it hardens and protects the meat from drying out unduly. Among domestic animals it is most prominent in sheep.

Fertility   In animal husbandry, the ability to produce normal young. Breeding capacity.

Fetus   The unborn young in the later stages of development.

Fiber   That portion of plant materials which comprise the cell walls and consist principally of cellulose, hemicellulose, and lignin.

Fibrotic (fibrosis)   A condition marked by the presence in the mammary gland of interstitial fibrous tissue resulting from mastitis.

Fill   The contents of the digestive tract.

Filly   A young female horse.

Finish   Refers to fatness. Highly finished means very fat.

Finishing   In beef cattle production, feeding thrifty calves or well-grown-out older cattle, mainly yearlings, a ration that is moderately high to very high in energy until they are sufficiently conditioned to yield a carcass with enough intramuscular fat to be palatable and acceptable to the consumer. The term also applies to swine and sheep.

First-calf heifer   A beef or dairy heifer following the birth of her first calf. At this time the heifer would usually be between 2 and 3 yr of age.

Fistula   A permanent opening made in the body wall of an animal through which there is communication between the animal's interior and the external environment. An example is a rumen fistula—a permanent opening created surgically between the interior of the rumen and the outside.

Fleece   The entire coat of wool as it comes from the sheep or while still on the live animal.

Flock (or **band**)   Refers to the total number of sheep under one management. Flock usually denotes small numbers on the farm, whereas "band" is used to designate large numbers on the range.

Fluid Milk   Milk commonly marketed as fresh liquid milks and creams.

Flushing   The practice of feeding female animals especially well, just previous to the breeding season, in order to stimulate their reproductive organs to the maximum activity. This is thought to result in a higher percentage of conceptions.

Flush season   Time during the year, especially the spring and early summer, when maximum milk production is expected in dairy herds.

Foaling   Act of a pregnant female horse (mare) giving birth to young.

Fodder   Coarse food, such as corn stalks, fed to cattle or horses.

Food and Drug Administration (FDA)   An agency of the federal government, part of the Department of Health and Human Services, that is responsible for safeguarding the quality and safety of foods and drugs that are used by humans.

Forage   Roughage of high feeding value. Grasses and legumes fed at the proper stage of maturity and quality are forages.

Forb   Weedy or broad-leaved plants, in contrast to grasses, that serve as pasture for animals.

Fore flank   The underside of the animal's body immediately behind the point at which the foreleg is attached to the body.

Forehead   Area between the eyes and immediately below the poll.

Fore udder   That part of the cow's udder which extends from the front teats to the point at which the udder is attached to the cow's body.

Fore udder attachment   That point on the cow's body at which the fore udder is attached to the body.

Foxhead   A device used to hold in an expelled (prolapsed) uterus after it has been cleaned and replaced in the ewe. It is made of ¼-in wire in the shape of a foxhead.

Free-martin   A heifer, usually sterile, born twin with a bull.

Free stall housing   A housing system for dairy cattle in which the barn is constructed with a series of stalls (one per cow) and in which the cow can choose in which stall she wishes to rest.

Fresh cow A cow that has recently given birth to a calf.

Freshen Commonly used to designate the act of calving (parturition). To give birth to a calf and concurrently initiate lactation.

Full feeding Providing animals with the maximum amount of feed needed to satisfy the functions for which they are being fed.

Fuzzy lamb A lamb covered with hair instead of wool. Such lambs are undesirable and often die at an early age or are unthrifty.

G

Gametes Reproductive cells. Sperm produced by the male and ova (or eggs) by the female.

Gene The hereditary materials carried in the chromosomes.

Genotype The complete genetic makeup of an animal.

Generation The time from the birth of one generation to the birth of the next.

Gestation The period in the female's life during which the unborn young is carried. The time between conception and calving.

Get of sire(livestock shows)   Consists of four animals, either sex or any age, the offspring of one sire.

Gilt A female swine that has not produced pigs and has not reached an evident stage of pregnancy.

Gimp (gimpy)   A lame animal.

Girth The measure (or distance) around the body of an animal.

Gland An organ for secreting a substance to be used in or eliminated from the body.

Globulin A protein that is not soluble in water but soluble in dilute salt solution. A protein found in blood and in milk.

Globules Small, round particles. Fat in milk is found in the form of globules.

Goad A pointed stick or rod used to urge on animals.

Goitrogenic area An area or section of the country in which the soil is deficient in iodine, thus causing feeds grown on the soils to be iodine-deficient.

Gonads The glands that produce the reproductive cells. The testis in the male and the ovary in the female.

Grade An animal that is not a purebred but that carries more than 50% of its genes from a recognized breed.

Grading up The continued use of purebred sires of the same breed in a grade herd or flock.

Grafting Inducing a newly lactating ewe without a lamb to accept a lamb not her own.

Grand champion The most outstanding animal of either sex in a livestock show from a type (conformation) standpoint.

Grass A plant commonly used as forage for animals. This term includes cereal grains (wheat, oats, rye, barley, corn) and pasture grasses.

Green chop   A system of forage harvesting where the green herbage is cut and chopped in the field, then fed to animals in confinement.

Grooming   The brushing, currying, and cleaning of an animal for appearance and sanitary or other reasons.

Group housing   A system of loose housing of dairy cows where the cows rest unstanchioned in a common, bedded area on a manure pack.

Growing-fattening   Refers to that period in swine production from weaning (about 8 weeks of age) to market weight of about 225 lb.

Guard rail   A rail attached to the walls in a farrowing pen that prevent the sow from laying on and crushing newly born pigs. The rail is usually 8 to 10 in from the floor and extends 8 to 12 in from the walls.

Gummer   A sheep with its incisor teeth missing.

Gutter cleaner   A mechanism in a barn designed to clean manure from the gutter.

H

Ham   The rear quarter of a pig or of a pork carcass.

Hammer mill   An item of equipment for grinding feed in which the grinding process is performed by several rows of thin steel hammers revolving at high speed.

Hand mating   Bringing a female to a male for service (breeding). After mating, the female is removed from the area where the male is located.

Hand milking   The manual milking of an animal, as opposed to mechanical milking.

Haploid   A term used in genetics to indicate the chromosome number occurring in the gametes. These cells, the sperm and the ovum, have one member of each chromosome pair. One-half the diploid number.

Hardening off   The practice of fasting fattening hogs for a short time when changing them from a diet containing unsaturated fats to one containing saturated fats. This causes a firmer body fat in the pork carcass.

Hay   Dried forage (e.g., grasses, alfalfa, clovers) used for feeding farm animals.

Haylage   Forage ensiled at a relatively low moisture content (usually 40 to 50%).

Heart girth   The circumference of an animal at a point immediately back of the shoulders.

Heat (as related to reproduction)   Sexual excitement by females. The period during which the female will accept service by the male. Technically known as estrus.

Heat-mount detector   A 2 × 4½ inch fabric base to which is attached a white plastic capsule containing a tube with red dye. The detector is glued to the rump of a cow or heifer in which the owner wishes to detect estrus. When the animal is in estrus, and is mounted by other females, the dye is released from the tube and the capsule turns red.

Heavy ewes   Female sheep near parturition as indicated by their full sides and bellies and looseness around the genital parts.

Heifer   A bovine female less than 3 yr of age who has not borne a calf. Young cows with their first calves are called first-calf heifers.

Herbivorous animals   Animals that habitually rely upon plants and plant products for their food.

Herd   A number of cattle, or other large animals, assembled together in a definite group.

Herdmates (stablemates or contemporaries)   A term used when comparing records of a sire's daughters with those of their nonpaternal herdmates, both being milked during the same season of the year under the same conditions.

Hereditary   A trait capable of being transmitted from parent to offspring.

Heritability   A term used by animal breeders to indicate what proportion (fraction) of the variation in a trait is due to differences in genetic value rather than to environmental factors.

Hermaphrodite   An individual having both male and female reproductive organs.

Herringbone milking parlor   A type of milking parlor widely used by commercial dairymen. In this type of parlor, cows are aligned in close proximity with the posterior one-third of their bodies exposed. The alignment of the cows is in a herringbone pattern. A double-six herringbone parlor means six cows on either side of the milking pit.

Heterosis   The increased stimulus for growth, vigor, or performance often shown by a crossbred individual. Hybrid vigor.

Heterozygous   The genetic situation where paired chromosomes carry dissimilar genes affecting a specific trait.

Hide brand   A permanent identification marking on the hide of an animal made with a hot iron, branding fluid, or by freezing the skin for a few seconds.

High-moisture grain   Cereal grain, most commonly corn, stored at a moisture content of 22 to 40% in a silo.

High-moisture silage   Silage containing 70% or more moisture.

Hobble   Tying the front legs of an animal altogether, or the hind legs to a rope run between the front legs and over the shoulder, to prevent kicking. Also a fetter for a cow or other animal.

Hock   The tarsal joint in the hind leg of a cow or other quadruped.

Hog sprays and wallows   Use of fine sprays of water or muddy areas where pigs can recline to protect themselves against high temperatures.

Holding area   A pen that accommodates one to four groups (one group fills the milking parlor) of cows just prior to milking.

Holocellulose   Total structural carbohydrate in a plant. It excludes soluble sugars, starch, and pectins.

Homeotherm   A "warm-blooded" animal. An animal that maintains its characteristic body temperature even though environmental temperature varies.

Homogenize   To force a substance through a small opening under pressure. Milk is homogenized to break the fat globules into much smaller units than occur in the natural state.

Homozygous   A term used to describe the genetic situation where paired chromosomes carry similar genes affecting a specific trait.

Honeycomb (reticulum)   In animal terminology, the second compartment of the ruminant stomach. The lining of this compartment resembles a honeycomb in structure.

Hooks   In cattle, refers to the hip bones. The point at which the femur articulates with the sacrum.

Hormone   A chemical messenger secreted by ductless glands in the body and which exerts a profound effect on physiological function. Examples are thyroxin, estrogen, cortisone, etc.

Hothouse lambs   Lambs born in fall or early winter and marketed to a special trade when 9 to 16 weeks of age or from Christmas to May.

Husbandman, husbandry   Terms applied to a farmer and his business (i.e., raising crops and/or livestock).

Hybrid   In genetics, an individual resulting from the mating of individuals belonging to different genotypes or of different genetic makeup.

Hydroponics   Growing of grains sprouted in chambers under conditions of controlled temperature and humidity to provide a source of green feed.

Hypocalcemia   Below normal concentration of calcium in the blood. Hypocalcemia is typical of parturient paresis (milk fever).

I

Immune   Exempt from disease.

Immunity   A condition in which the animal becomes resistant to disease either through the use of vaccines or serums or through the production of antibodies resulting from having had the disease.

Impotence   The loss of breeding power; incapacity for sexual intercourse; incapability of serving as a breeder.

Impregnate   To make pregnant; to fertilize.

Inanition   Exhausted from lack of food or nonassimilation; state of fasting.

Inbreeding   Mating of animals more closely related than the average relationship found in the population.

Infectious disease   A disease that is transmissible from one animal to another.

Insemination   The deposition of semen in the female reproductive organs, usually the cervix or the uterus.

Inseminator   The skilled technician who is qualified to perform all the duties involved in artificial insemination.

Instinct   Inborn behavior.

Integrated Reproduction Management   Management of animals in such a way as to enhance reproductive efficiency. This involves attention to several management factors including nutrition, health, etc.

Internal parasites   Parasites found within the animal's body, most commonly within the gastrointestinal tract but occasionally in the lungs, liver, kidneys, and muscles.

In vitro   Within an artificial environment, as within a test tube.

In vivo   Within the living body.

Involution   A decline in size or activity of tissues and/or organs. For example, the mammary gland tissues normally involute with advancing lactation. The uterus involutes within a few days or weeks after a female gives birth to young.

Iodinated casein (thyroprotein)   A product produced when casein, a milk protein, is treated with iodine. The resulting product possesses thyroactivity and is sometimes fed to milk cows to stimulate milk production.

Isolation stalls   Pens for keeping animals separated from the rest of the herd.

J

Jail   A small pen large enough to hold only one ewe and her offspring. Also called a "jug."

Jowl   That portion of the head of a pig immediately to the rear of the lower jaw.

Judging   Comparing several animals on the basis of appearance.

Junior pig (swine shows)   Farrowed on or after 15 March of the year shown.

Junior yearling (cattle shows)   An animal born between 1 January and 30 June of the previous year.

Junior yearling (swine shows)   Farrowed on or after 1 March and before 1 September of the previous year.

K

Kid   A young goat up to its first birthday.

Kidding   Parturition in the goat; the act of giving birth to young.

Kidding (rutting) period   Term designating when goats are in heat.

L

Lachrymation   The act of tearing; secreting and conveying tears.

Lactalbumin   A component of milk protein.

Lactation   Production of milk by the mammary gland.

Lactation period   The number of days a cow secretes milk following each parturition.

Lactic acid   An acid formed in the souring of milk. Also produced when forage carbohydrates are fermented in silage production.

Lactogenic   Stimulating the secretion of milk.

Lactogenic hormone   The pituitary hormone that stimulates the secretion of milk. Same as prolactin.

Lactose   The principal sugar of the milk of all mammals. A disaccharide composed of one molecule of glucose and one molecule of galactose.

Lagoon   A small body of water or pond into which liquid manure is discharged and is subsequently digested by bacterial action.

Lamb   A sheep under one year of age. If a lamb has lost any of its temporary teeth, it would not be classed as a lamb. The same term is used for the meat of a young sheep.

Lambing   Giving birth to lambs.

Lambing pen (lambing jug)   A small pen, about 4 ft square, in which the ewe is placed immediately before lambing or, if lambing has occurred elsewhere, immediately after lambing.

Lard   The fat rendered (melted out) from fresh, fatty pork tissue.

Lariat   A long, thin rope of hemp or strips of hide with a running noose, used for catching cattle, horses, etc.; a lasso.

Late lambs   Lambs born after the normal lambing time for a particular area has passed.

Lead feeding (challenge feeding)   The practice of increasing grain fed to cows to a level equal to 1.0 to 1.5% of body weight beginning about 3 weeks prior to the predicted calving date. After calving, feeding is continued at a high energy level until maximum economic production is reached.

Lead poisoning   Animal poisoning due to the ingestion of lead, usually from lead-containing paints.

Lethal characters   Presence of hereditary factors in the germ plasm that produce an effect so serious as to cause the death of the individual either at birth or later in life.

Leucocyte A white blood corpuscle.

Libido Sexual desire or instinct.

Lignin A group of complex chemical substances found in the fibrous parts of plants. Lignin is essentially indigestible.

Limited feeding A cattle feedlot term meaning not giving the animal all the feed it will consume.

Linebreeding A form of inbreeding in which an attempt is made to concentrate the inheritance of some ancestor in the pedigree.

Lipase A fat-splitting enzyme. When found at high levels in milk or milk products, lipase causes rancidity.

Lipid Any of a group of biologically important substances that are insoluble in water and soluble in fat solvents such as ether, chloroform, and benzene. Principal chemical components of lipids are fatty acids and glycerol.

Liquid feeding A term common to swine husbandry that refers to the practice of mixing dry feed with a fluid, usually water or skim milk, before feeding and offering this high moisture mixture to the animal. Also called slop feeding.

Litter The offspring of a litter-bearing female delivered at one parturition.

Live backfat probe A system for determining the amount of backfat in a live pig, which consists of making incisions in the back of the pig at appropriate points and then measuring the thickness of the backfat at each point with a metal ruler.

Live weight In meat animals, the weight before slaughtering.

Loading chute A ramp by which animals may be loaded onto a truck.

Loin That portion of the animal's back extending from just in front of the hip bones forward to an area about even with the most posterior rib.

Loin eye area An indicator of muscling in a carcass. In swine, the loin eye muscle is measured between the 10th and 11th ribs. In cattle it is measured at the 12th rib.

Longevity Refers to cows that "wear well" (i.e., that stay in the herd for a long time).

Long feed Coarse or unchopped feed such as hay, in contrast to short, or chopped, feed.

Loose housing A housing system for cattle whereby animals are untied and unstanchioned in a shelter and are free to move at will from the inside to the outside and from a feeding area to a resting area.

Low-moisture silage Silage that usually contains between 40 and 50% moisture.

Low test   In dairy cattle terminology, this term refers to cows with a relatively low percent of fat in their milk.

Lucerne (alfalfa)   A legume of high feeding value for ruminants.

Lunger   Used in reference to a sheep with a chronic respiratory disease that produces difficult breathing.

M

Machine stripping   Pulling down on the teat cups of the milking machine at the conclusion of milking to remove residual milk.

Maintenance requirements (maintenance ration)   The nutrients required by an animal that is doing no work and yielding no product.

Mammary gland   Complex glands, present in the female, whose purpose is to produce milk for the nourishment of the young.

Mammary system in cows   That part of a cow which includes the udder, udder attachments, milk veins, and teats.

Mammary veins (milk veins)   Blood veins visible on the underside of the cow and extending forward from the fore udder.

Managed milking   A systematic, efficient way of organizing the milking of a herd of dairy cows.

Management   Managing, handling, controlling, or directing a resource or integrating several resources.

Management systems   Methods of systematically organizing information to assist in making more effective management decisions.

Manure   Excrement of animals.

Manyplies (omasum)   Common name for the third stomach compartment in the bovine.

Marbling   Fat deposits in the lean tissue in meat.

Mare   A mature female horse.

Market class   The use to which animals are put. Market class is determined by all of those factors affecting the use and value of the animal, except the final grade.

Market grade   A measure of how well the animal fulfills the requirements for a class.

Marketing   The act of selling or purchasing, as in a market.

Marketing agreement, milk   A contract issued by the Secretary of Agriculture specifying that handlers of milk will pay producers in a given milk shed a certain minimum schedule of prices for their milk.

Marking   A term applied to docking and castrating lambs (also called "cutting"); also branding or marking for identification.

Mastectomy   Removal of the mammary gland.

Mastication   The process of chewing food previous to swallowing.

Mastitis   Inflammation of the mammary gland.

Maternity stall   A pen of 100 to 120 sq ft in which pregnant cows are placed from a day or two before to a day or two after calving.

Mature equivalent (ME)   In dairy terminology, a production record of an immature cow adjusted by means of factors to a mature basis. Similarly, weights of offspring of young beef cows, ewes, and sows are often adjusted to a mature basis.

Meat type hog   An intermediate type hog with an anatomical structure about halfway between the short, chuffy type hog and the very large, long hog. Produces a high percentage of lean pork cuts.

Meiosis   The process in gametogenesis in which the chromosomes are reduced from the paired (diploid) to the single (haploid) number.

Mendel's Laws   The basic laws of inheritance discovered by Gregor Mendel, an Austrian monk, in breeding experiments with peas.

Metabolic disease   A noninfectious disease caused by a physiological derangement (e.g., milk fever, ketosis, and pregnancy disease).

Metabolic weight   The weight of an animal raised to the three-quarter power ($W^{.75}$).

Metabolism   Chemical changes proceeding continually in living cells, which release energy and build up protoplasm.

Metritis   An inflammation of the uterus.

Microbiology   Science dealing with the study of microscopic organisms in both the plant and animal kingdoms.

Microflora   Microbes characteristic of a region, such as the bacteria and protozoa populating the rumen.

Milk ducts.   Passages through which the milk moves from the time of its synthesis in the alveoli to its withdrawal from the udder.

Milk fat (butterfat)   The fatty material found in milk. A characteristic of milk fat is the presence of low molecular weight, saturated fatty acids.

Milking machine   A mechanical device used to extract milk from cows. A milking machine relies on vacuum to remove milk via the teats from the udder.

Milking parlor   A structure in which cows are milked. Cows are in the milking parlor only during the milking period and, in some systems, receive no feed while in the parlor.

Milk letdown   The squeezing of milk out of the udder tissue into the gland and teat cisterns.

Milk meter   A device used in milking parlors to measure the rate of flow and thus the pounds of milk produced by a cow.

Milk only record   A type of unofficial DHIA production testing record made under the National Cooperative Dairy Herd Improvement Program in which milk weights are recorded at the farm once monthly but in which no milk composition analyses are made.

Milk replacer   A feed material for young animals which has many of the nutritive characteristics of milk, is fed in a fluid form, and contains an appreciable level of nonfat dry milk solids.

Milk well   Opening in the cow's abdomen through which the mammary vein passes on its way to the heart from the udder.

Minor elements   Essential elements used in very small amounts by animals. Sometimes called trace elements.

Miscarriage   The failure of a female to carry her fetus to a normal birth.

Mitosis   Cell division in which there is a first dividing of the nucleus followed by the dividing of the cytoplasm.

Mixed feed   A concentrate feed consisting of several constituents such as cereal grains, cereal grain by-products, or protein supplements.

Monogastric   Having only one stomach or only one compartment in the stomach. Examples of monogastric animals are swine, mink, and rabbits.

Monoparous   A term designating animals that usually produce only one offspring at each pregnancy. Horses and cattle are monoparous.

Monorchid   A male animal that has only one testicle in the scrotum.

Mother up   Commonly used in two senses: (1) Observation of newly lambed ewe in a jug to make sure that she accepts her lamb, that she is lactating, and that the lamb is on his feet and nursing. (2) Allowing a period of time for ewes and young lambs to find each other (mother up) after such procedures as marking, docking, and vaccinating. The term is also often used with beef cattle.

Motility   The amount of activity (sperm movement) in a semen sample observed under a microscope.

Mount   To climb onto, as demonstrated be females riding other females in heat and by males in natural service.

Mouthing   Determining the age of animals by examination of their teeth.

Mow drying   A system for reducing the moisture content of harvested hay once it is the mow. The system normally includes tunnels in the hay mow or stack through which air (heated or unheated) passes until the hay is suitably dried.

Muley   A polled cow.

Multiparous   Refers to a female that has given birth to several young.

Mummified fetus   A shriveled fetus remaining in the uterus rather than being absorbed or expelled.

Mutation   A process by which a sudden genetic variation occurs in the genetic composition of an animal and subsequently breeds true.

Mutton   Meat from a mature sheep.

Muzzle   The nose and projecting jaws of an animal such as a cow.

Mycotoxicoses   A toxic condition exhibited in animals and caused by excess consumption of toxins produced by molds and fungi.

N

National Association of Animal Breeders The principal organization of artificial breeding associations in the United States.

National Milk Producers Federation A national association of milk marketing cooperatives devoting attention to the various problems of the dairy industry with emphasis on marketing of milk and dairy products.

National Research Council A division of the National Academy of Science responsible for producing publications on the nutrient requirements of domestic animals.

Natural service Refers to natural mating in farm animals, as opposed to artificial insemination.

National Swine Improvement Federation An organization of swine producers and others interested in the genetic improvement of swine that recommends guidelines for on-farm swine testing.

Neck strap A leather thong fastened around the neck of a cow or calf for identification and/or for fastening the animal.

Needle teeth The small, tusklike teeth on each side of the upper and lower jaws in swine. In good swine management, the tops of these teeth are removed shortly after birth to avoid damage to the sow's udder by the nursing pig.

Negative balance A term used in animal nutrition or physiology that describes a situation where an animal is excreting or secreting more of a specific nutrient than it is ingesting.

Neonatal Pertaining to a newborn.

Neonate A newborn infant.

Nick In livestock breeding, an animal genetically superior to either parent.

Nipple pail A pail to which a nipple is attached for feeding milk to a young calf. This feeding system simulates the nursing act.

Nitrogen-free extract (NFE) That part of the carbohydrate fraction of a feeding stuff which is most readily digested. Consists principally of starch, sugars, and non-nitrogenous organic acids.

Noninfectious disease A disease not transmissible from one animal to another. Examples are metabolic and nutritional diseases and diseases caused by poisonous plants and toxicants.

Nonreturns The conventional method of measuring "fertility" in artificial breeding. If, after an original insemination, a cow is not reported for reinsemination within a certain designated period, she is considered a "nonreturn" and is assumed to be pregnant.

NPN Nonprotein nitrogen (e.g., urea).

Nulliparous A term referring to a female that has not given birth to young.

Nurse cow   A dairy cow used to supply milk for nursing calves in addition to or in place of her own.

Nursery   In swine husbandry, a structure specially designed to raise young pigs between 3 and 8 weeks of age. A swine nursery is usually in an environmentally controlled building.

Nutrient   A term applied to any food constituent, or group of food constituents of the same general chemical composition, that contributes to the support of animal life.

Nutrient toxicity diseases   Diseases caused by excess consumption of essential nutrients.

Nutritional diseases   Diseases caused by real or induced deficiencies of essential nutrients.

Nymphomania   An abnormal reproductive condition in cattle in which the female is in more or less constant estrus.

O

Offal   The parts of a butchered animal that are removed in dressing it. Generally has reference to the inedible parts.

Off feed   A term often used in reference to the loss of appetite in farm animals.

Offspring   That which springs from an animal or plant as an individual reproducing its kind; progeny; issue.

Old crop lamb (or **old cropper**)   A feeder lamb that did not reach market weight and finish during its first season and must be held on feed into the winter or early spring.

On full feed   Refers to feeding systems where animals receive all the feed they will consume.

Open   In livestock breeding, an animal not pregnant.

Open-faced   Sheep with little or no wool on the face, especially around the eyes.

Open formula feed   Mixed concentrate feed in which the ingredient amounts in the feed formula are available to the purchaser.

Openness   A physical quality in dairy cattle associated with length of body and width and length of rib.

Orchidectomy   The surgical removal of one or both testes; castration.

Organoleptic properties   Refers to properties detected by the sense of smell and taste.

Orphan   A young animal that cannot be nursed by its mother due to her sickness or death.

Out-crossing   The mating of unrelated individuals of the same breed.

Ovariectomy   The surgical removal of an ovary.

Ovary   Paired structures in the female that produce both the ova (eggs) and female sex hormones.

Overconditioned   A condition indicated by excess fleshing and patchy fat deposits, especially over the shoulders, back, and thighs.

Overshot jaws   A condition where the upper jaw is longer than the lower jaw, preventing the teeth from meshing properly (malocclusion). The opposite condition is known as undershot.

Ovulation   Extrusion of the ovum from the ovary. Rupture of the mature ovarian follicle with release of the contained ovum.

Ovum   The egg or female sexual cell produced in the ovary.

Ovum transplantation   See *embryo transplant.*

Owner-sampler records   A production testing program in DHIA whereby the owner himself, rather than a supervisor, weighs and samples the milk.

Oxytocin   A hormone secreted by the pituitary gland that causes a cow to let down her milk when the proper stimulus is applied.

P

Palatability   In animal feeding, refers to the avidity with which an animal selects from among several feeds for ultimate consumption.

Pale soft exudative (PSE) pork   Pork that is soft, pale, and loose-textured, with little or no marbling. An inherited trait closely related to the porcine stress syndrome (PSS).

Pancreas   A complex organ associated with the digestive tract and producing the pancreatic juice, which contains several digestive enzymes. Also produces insulin.

Parathyroid glands   Small endocrine glands located near or imbedded in the thyroid gland. These glands secrete a hormone that is important in regulating the calcium content of the blood.

Parenteral   Administration of medication by injection (e.g., subcutaneous, intramuscular, intrasternal, intravenous), not through the digestive tract.

Parous   A term referring to females having produced one or more young.

Parturition   The act of the female in giving birth to young.

Pastern   That part of the animal's fore and rear legs located between the dew claws and the hoof.

Pasture   Plants grown as feed for grazing animals. Also, to feed cattle and other livestock on pasture.

Pasture rotation   The rotation of animals from one pasture to another so that some pasture areas have no livestock on them in certain periods.

Patchiness   A term indicating the presence of fatty deposits about the tailhead and other parts of dairy and beef cattle.

Pathogen   Any microorganism (bacteria, viruses, yeasts, molds, helminths) that produces disease.

Pathologist A specialist in the study or treatment of disease.

Paunch (rumen) First compartment of the stomach of the bovine.

Pearson square method An arithmetic method used in standardizing milk and in balancing rations.

Pedigree A list or table showing an animal's line of ancestors and their relationships.

Pellets A physical form of feed that consists of compressing ground roughage or concentrate into a relatively dense tablet.

Pelt The skin and wool of a sheep; the skin and attached hair (fur) of any fur-bearing animal.

Pendulous udder Loosely attached mammary gland.

Pepsin The principal enzyme found in the gastric juice secreted by the stomach. Converts proteins into simpler compounds such as peptones and proteoses.

Performance testing Merit testing of individual animals, usually for rate of weight gain, feed efficiency, milk production, or other performance trait.

Persistency The quality of being persistent, as in the ability of lactating animals to maintain milk production over a period of time.

Pesticides Chemical such as insecticides, fungicides, and herbicides used in crop and animal production to control insects, weeds, etc.

pH A numerical value that expresses degree of acidity or alkalinity of a material as the negative logarithm (to the base 10) of the hydrogen ion concentration.

Phagocytes White blood cells that destroy bacteria in the bloodstream.

Phenotype A genetic term that refers to an animal's appearance or performance characteristics in contrast to the animal's genetic composition.

Physiology That branch of the science of biology which deals with the normal, vital functions of various part of the living animal.

Pica An appetite for materials not usually considered to be food such as is observed in phosphorus-deficient animals; a depraved appetite.

Picnic shoulder The lower portion of the shoulder of the pork carcass below the Boston butt and above the knee joint.

Pig brooders Pens or enclosures with a source of heat in which very young pigs (2 to 3 days old) are placed to prevent their exposure to cold.

Pigging See *farrowing*.

Pinning Collection of dung around the vent of very young lambs that has dried to the point of interfering with normal bowel movements.

Pins The prominences located on either side of the tail setting in cattle. Anatomically they represent the most posterior portion of the sacrum.

Pipeline milking   A milking system where the milk withdrawn from the cow by a milking machine moves from the cow to a stainless steel or glass pipeline and ultimately to a storage tank.

Pituitary gland   A small endocrine gland producing a number of hormones vital to the functions of the animal. This gland is located at the base of the brain.

Placebo   An inactive substance or preparation used as a positive control in some research studies as a method of examining the value of some potentially active substance that may have some value in medication.

Placenta   The vascular structure by which the fetus is nourished in the uterus.

Plain   A term suggesting general inferiority; coarse; lacking the desired quality or breed character.

Plum Island Animal Disease Center   A federal research center of USDA's Agricultural Research Service located on a small island at the east end of Long Island, New York. Exotic animal diseases are studied at this location.

Point of shoulder   The most anterior portion of the shoulder in cattle. Anatomically the most anterior portion of the shoulder blade or scapula.

Poisonous plants   Plants containing substances causing toxic conditions, and sometimes death, when consumed in sufficiently large amounts by grazing animals.

Poll   The top of the head between the horns of a bovine, sheep, or goat.

Polled   Naturally hornless cattle, sheep, or goats.

Polygastric   Possessing more than one stomach compartment, as do the cow and other ruminants.

Polysaccharide   A compound composed of a number of simple sugar units joined together forming a complex unit of high molecular weight. Examples are starch, glycogen, and cellulose.

Porcine   Pertaining to swine.

Pork   The meat that comes from swine.

Post-legged   A condition in which the hind legs are too straight, so that the springy quality of the hock and pastern is lost.

Postpartum   Occurring after birth of the offspring.

Potent   Having the power of procreation.

Ppm (parts per million)   An expression of the concentration of a material in a solution. Equals milligrams per kilogram or microliters per liter.

Predicted difference   A term used in connection with dairy cattle breeding that indicates the amount by which a bull's daughters should outproduce (or underproduce) their breed average herd mates on a mature equivalent basis.

Predisposing causes of diseases   Factors that act to cause the animal body to be especially susceptible to disease.

Pregnancy testing   Refers to physical examination of the female or chemical testing of her body fluids to determine whether or not she is pregnant.

Pregnant   Condition of being with young.

Premature birth   Expulsion of the young before full term.

Prepartum   Occurring before birth of the offspring.

Prepartum milking   The practice of milking cows before parturition.

Prepotency   The ability of an animal to stamp its own characteristics on its offspring.

Produce of dam(cattle shows)   Consists of two animals, either sex or any age, the produce of one cow.

Produce of dam(swine shows)   Consists of four animals, either sex or any age, the produce of one sow.

Production testing   Genetic testing of animals. Includes both performance and progeny testing.

Progeny   Offspring of animals.

Progeny testing   Genetic testing of farm animals based on the performance or production of the offspring of a male or female.

Progestagens   Compounds that mimic the hormone progesterone, the sex hormone produced by the corpus luteum of the ovary.

Progesterone   A female sex hormone secreted by the corpus luteum of the ovary.

Prolapsed uterus   A condition in which the uterus is partially or completely turned inside out and protrudes to the outside of the body. This condition occasionally occurs following parturition.

Prolificacy   Ability to give birth to a large number of young. This term can refer to individual animals, groups of animals, or species.

Prophylactic   A preventive, preservative, or precautionary measure that tends to ward off disease or an undesirable condition.

Proprietary feeds   Ready-mixed feeds prepared according to a company-controlled formula and sold with a guaranteed chemical analysis.

Propionic acid   A three-carbon organic acid formed by a number of biological systems including the action of bacteria on dietary carbohydrates in the rumen of the bovine.

Prostaglandins   Hormonelike substances found in every body cell and tissue and believed to play a key role in regulating cellular metabolism.

Protein   A nitrogenous organic constituent of food or feed manufactured in plant or animal tissue. An essential nutrient in animal nutrition.

Proven sire   A dairy bull whose genetic transmitting ability has been measured by comparing the milk production performance of his daughters with that of the daughter's dam and/or herd mates under similar conditions.

Proximate analysis   A chemical analysis that represents the gross composition of a feed. The feed components identified by this procedure include nitrogen (crude protein), nitrogen-free extract, ether extract, ash, and dry matter.

Ptyalin   A starch-splitting enzyme found in the saliva.

Puberty   The period of life at which the reproductive organs first became functional. This is characterized by estrus and ovulation in the female and semen production in the male.

Purebred   Any animal that traces back through all its lines to the foundation stock of the breed it represents. In some breeds, animals resulting from four or five generations of crossing to purebred sires are recognized as purebreds.

Q

Quality   In animal husbandry, refers to the relative fineness of structure of an animal. It is indicated by fine hair, loose and pliable skin, an even covering of firm, mellow flesh, fine but strong bone, a clean-cut head that is free from coarseness, and a stylish appearance.

R

Rack   A wooden or metal frame in which fodder is placed for cattle, horses, or sheep. Also, a certain gait of a horse or other quadruped. Also, the area of the body, or especially the carcass, containing the ribs but excluding shoulder, breast, and loin.

Ram   An uncastrated male sheep.

Range (in agriculture)   A term generally applied to the open grasslands of the plains region of North America, where cattle, sheep, and horses feed.

Range-bred   Bred and reared on the range; accustomed to living in the open country as range-grown cattle.

Rangy type hog   A hog characterized by a relatively long body and long legs.

Ranting   A behavioral activity of some boars that consists of their pacing back and forth along the fence, chopping their jaws, and slobbering.

Rate of passage   the time taken by undigested residues from a given meal to reach the feces.

Ration   the amount of feed supplied to an animal for a definite period, usually for a day.

Reactor   An animal giving a positive reaction to such health status tests as tuberculin or brucellosis tests.

Rear flank   The fold of skin on the underside of an animal's body immediately in front of the point where the rear leg meets the barrel.

Rear udder   All of the cow's udder from a point halfway between the front and rear teats to the most posterior portion of the udder.

Rear udder attachment   The folds of skin that attach the rear udder to the animal's body.

Recessive character   A term used in genetics for describing a trait or character of an animal that is expressed only when the gene responsible for it is present in both members of a chromosome pair.

Reference sire   Sire whose progeny performance is used as a reference for evaluating progeny performance of test sires. Generally, a reference sire should have 100 or more progeny evaluated in 10 or more herds compared to progeny of five or more other sires. Reference sire programs are used mostly in beef sire evaluation.

Refractory   Resistant to treatment or cure; unresponsive to stimulus.

Regurgitate   To cast up undigested food from the stomach to the mouth as done by ruminants.

Registered (livestock)   An animal whose name has been recorded and numbered by a recognized breed association.

Repeatability   the tendency of an animal to repeat its performance (e.g., a dairy cow in successive lactations, a sheep in successive wool clips, etc). In dairy bulls, this term is used to indicate certainty with which predicted difference reflects true transmitting ability.

Reproductive cycle   The sexual cycle of the nonpregnant female. Characterized by the occurrence of estrus (heat) at regular intervals (e.g., in the cow, every 21 days).

Respiration calorimeter   An apparatus for measuring the gaseous exchange between an animal and the surrounding atmosphere. With this apparatus it is possible to measure the energy required for the various body functions and activities and the energy content of feeds.

Retail cuts   Subdivisions of the carcass of any meat-producing animal that are commonly sold over the counter in retail meat markets.

Retained placenta   Placental membranes not expelled at parturition.

Reticulum (honeycomb)   Second compartment of the bovine stomach.

Rib eye area   Cross section area of large muscle lying in angle of the last rib and vertebra. Indicative of bred-in carcass muscling.

Ride   To be transported in a mounted position, as one cow mounted on another during estrus.

Ringing   The practice of placing a ring or rings in the snout of a pig to prevent rooting.

Roan   Describes the color of an animal's hair coat. The basic hair color of a roan is bay, chestnut, red, black, or brown color, with some white hairs thickly interspersed.

Roaring buck   Ram with a laryngeal infection, usually fatal, that produces very difficult breathing. Also called "wheezer."

Roasters   A market term referring to fat, plump, suckling pigs, weighing 15–50 lb on foot.

Rolling herd average   The average milk production per herd per year based on the 12 mo just past.

Rotation grazing   A grazing system in which small fields or paddocks are fenced off with the fields being grazed in a definite sequence. During the grazing cycle, the fields remain ungrazed for a period of time, thus allowing the grass to regrow until the field is grazed again.

Roughage   Consists of pasture, silage, hay, or other dry fodder. Roughage is relatively low in energy and high in fiber.

Roughs   A swine marketing term referring to coarse, rough hogs lacking in condition.

Rule of thumb   A general rule that may be applied in arriving at a decision.

Rumen (paunch)   The first, and largest, of the four stomach compartments of the ruminant.

Ruminants   Animals that ruminate, or "chew the cud." Characterized by a several-compartmented stomach in which the food consumed undergoes bacterial fermentation before moving to the "true" stomach.

Rumination ("chewing the cud")   The process of regurgitating and rechewing food; unique to ruminants.

Rump   That portion of the topside of an animal from the hip bones to the tail.

Runt   Any animal that is small compared to the rest of its kind.

S

Scoured (scouring)   Wool washed to remove yolk, dirt, and other natural impurities.

Scours   Severe diarrhea.

Scrub   A domestic animal of mixed or unknown breeding that does not have the markings or characteristics of a recognized breed.

Scurs   Rounded portions of horn tissues attached to the skin at the horn pits of polled animals.

Seedstock herd   A herd maintained for the primary purpose of providing breeding stock for use in commercial herds.

Selection   Biologically, any process, natural or artificial, that permits certain individuals to leave a disproportionately large (or small) number of offspring in a group or population.

Selection index   In animal breeding, a single overall numerical value derived by weighting values for several traits based on heritability and economic importance. Index values are used in determining which animals are selected for breeding and which are culled.

Self-fed free-choice feeding   A feeding system commonly used with swine and often with finishing cattle in which feed is constantly available in feeders, and the pigs can consume feed whenever they wish.

Self-feeder   A small structure for feeding animals cafeteria-style. A self-feeder consists of a storage bin for holding several hundred pounds of feed and a trough into which feed descends by gravity and where animals can eat.

Self-sucking   A habit, sometimes acquired by cows, of sucking their own teats.

Semen   The viscid, whitish fluid produced in the male reproductive organs, which contains the spermatozoa needed to fertilize the ovum of the female.

Semen extenders   Fluids of special composition used to dilute semen prior to its use in artificial insemination.

Semen, frozen   Semen that is frozen and kept at $-79°$ C (dry ice) to $196°$ C (liquid nitrogen) prior to its use.

Senior pig (swine shows)   Pig farrowed on or after 1 September of the previous year and before 1 February of the year shown.

Senior spring pig   Pig farrowed on or after 1 February and before 15 March of the year shown.

Senior yearling (cattle shows)   In show ring terminology, an animal 1½ or 2 yr of age.

Senior yearling (swine shows)   Swine farrowed on or after 1 September of two years previous and before 1 March of the previous year.

Service   A term commonly used in animal breeding, denoting the mating of male to female.

Set-up   Act of turning a sheep on its side or into a sitting position as a method of restraining.

Sex chromosomes   Chromosomes that behave as if they were members of the same pair but which are morphologically different in the two sexes. They, or factors carried in them, wholly or partially determine the sex of an individual.

Sex-linked   A term applied to the units of inheritance (genes) located on the sex chromosomes or to the characters conditioned by them.

Shank   That portion of the foreleg between the knee and the pastern.

Shear   The act of removing fleece from the body of sheep.

Sheath The skin surrounding the penis of the male animal.

Shell or shelly Poor, debilitated animal, frequently so from chronic disease. Also "pelter," "scab," "skate."

Shy breeder A male or female of any domesticated livestock that has a low reproductive efficiency.

Shrink Loss in weight that occurs when animals are transported to market.

Sib In genetics, a brother or sister. Full sibs have both parents in common; half sibs have only one parent in common.

Sib test Use of records of a group of full or half sibs to estimate breeding value of an individual. In swine production, refers to the feeding and subsequent slaughter of litter mates to estimate carcass quality and genetic value of breeding stock.

Sickle-hocked Describes an animal having a crooked hock, which causes the lower part of the leg to be bent forward out of a desirable line.

Side That portion of the hog's anatomy lying below the loin and back and above the belly. Bacon is made from the side of the hog.

Silo A structure in which green plant material is stored for the subsequent fermentation, which causes silage formation. Successful fermentation requires exclusion of air from the ensiled mass.

Silo, bunker An aboveground, rectangular structure, for storing silage, typically with walls 6 to 10 feet high. Bunkers are usually 40 to 100 or more feet long.

Silo, tower An upright, cylindrical structure in which forage is ensiled. Most tower silos vary in height from 30 to 60 feet and in diameter from 10 to 20 feet.

Silo, trench A horizontal structure in which forage is ensiled. The word trench implies that a suitable sized opening is made in the ground. Such horizontal silos do not always have wood or concrete walls. Aboveground horizontal silos are called bunkers.

Silo unloaders Mechanical devices for taking silage out of a silo and transporting it to a forage wagon or to a feeding area.

Sire index A figure that is indicative of the transmitting ability of a sire. In dairy cattle this figure is expressed in terms of quantity of milk or a milk component such as fat or protein.

Skim milk Fluid milk from which the fat has been removed.

Slaughter pigs A market class of pigs destined for slaughter and weighing less than 120 lb.

Slink To abort; or, an aborted fetus.

Slop-feeding A system of feeding animals (usually pigs) that involves mixing dry feed with a fluid, usually water or skim milk. Thus the animal consumes its feed in the form of a slurry.

Snout   In pigs, the nose and anterior portion of the upper and lower jaws.

Soft pork   Pork carcasses containing fat with a low melting point and thus relatively soft at storage temperatures. Soft pork is formed when a pig consumes relatively large amounts of unsaturated fat from feeds such as soybeans or peanuts.

Soil Conservation Service (SCS)   An agency of the USDA charged with developing and carrying out a permanent national soil and water conservation program.

Soiling crops   Refers to fodder or forage crops that are cut green and fed to animals before drying or fermentation.

Solids not fat   All of the solids in milk except the milk fat. Includes protein, lactose, and minerals.

Somatic cell count   The number of leucocytes and other body cells found in 1 ml of milk. A high count indicates problems with mastitis in a cow or a herd of cows.

Sorting   A method of separating animals into two or more subgroups from one group. This is often accomplished by means of a narrow chute, with gates to direct animals into separate corrals or pens. The chute used is variously called "dodge chute," "separating chute," "sorting chute," or "cutting chute."

Sow   A female swine that shows evidence of having produced pigs or that is in an evident stage of pregnancy. A sow that is obviously pregnant is called a "piggy sow."

Sow productivity index   A numerical value representing the genetic worth of a sow based on the number of pigs farrowed, pigs alive at 21 days, and litter weight.

Spay   To remove the ovaries of a female animal.

Specific-pathogen free (SPF)   In swine husbandry, the production of breeding stock reared under very carefully controlled, sanitary conditions and sold to swine breeders to reduce the chance of disease transmittal to their herds.

Sperm cells (spermatozoa)   The reproductive cells produced by the male in the testes.

Sphincter   A ring-shaped muscle that closes an opening, such as the sphincter muscles in the lower end of a cow's teat.

Springer   A term commonly associated with female cattle showing signs of advanced pregnancy.

Stable fly   A fly resembling the ordinary house fly but able to pierce the skin and suck the blood of animals and, therefore, a great nuisance.

Stag   An unsexed male animal, castrated when mature or so far advanced toward maturity that masculinity is rather evident in the forequarters.

Stallion   A mature male horse.

Stall, tie   A resting and/or feeding stall in which the cow is fastened to a chain that connects to her neck strap. Permits more freedom than a stanchion.

Stanchion barn   A barn equipped with stanchions where cows are held during feeding, resting, and/or milking periods.

Staple   A lock or small sample of wool from a fleece.

Steer   A male bovine castrated before development of secondary sex characteristics.

Sterile   As applied to animals, unable to reproduce.

Stifle   The bulge of muscle at the point where the rear leg meets the barrel immediately above the rear flank.

Stillborn   Born lifeless; dead at birth.

Stocker cattle   Refers to young animals so fed and handled that growth and reasonably good condition are the main objectives. Animals are fed chiefly on pasture and roughage. These animals are subsequently purchased by cattle feeders for fattening.

Stomach   The organ of the body into which the esophagus empties and which, in turn, empties into the small intestine.

Stover   Fodder; mature cured stalks of grain from which seeds have been removed, such as stalks of corn without ears.

Strip   To milk dry (i.e., to remove the residual milk from the cow's udder at the end of the milking act).

Strip cup   A cup that is used as a cowside test for mastitis. The top of the cup contains a fine screen through which a stream of milk from the teat is passed at the onset of the milking act. Clots or clumps of milk solids indicate a mastitis problem.

Strippings   The last milk of a milking.

Stud   A unit or herd of selected animals kept for breeding purposes. Often referred to as a seedstock herd.

Substance and strength   Terms used in livestock judging to describe animals with large bones, a deep body, a wide chest floor, and sizable heart girth.

Succulence   A condition of plants characterized by juiciness, freshness, and tenderness, making them appetizing to animals.

Superfetation in animals   Conception during a pregnancy. In such a case, the female will be carrying two fetuses or litters of different ages at the same time.

Supernumerary teats   Extra teats (more than four) in a bovine. In good dairy herd management, supernumerary teats are removed at an early age before they become highly vascularized.

Superovulation   The production by the ovaries of more than the usual number of eggs at the time of estrus. Usually induced by hormone treatment.

Surprise tests   Tests that are part of dairy cow production testing programs. Surprise tests are scheduled by the various purebred dairy cattle breed associations when a cow's milk and/or fat production exceed certain set levels. Surprise tests include the entire herd.

Swine breeding crate   A cratelike structure in which the female hog is placed at the time of mating with the male. The purpose is to keep the female in a relatively fixed position during the mating act.

Symmetry (in livestock)   Perfection of proportion in build; harmony of all parts of an animal viewed as a whole with regard to the approved type of the breed it represents.

T

Tack room   Refers to a room, usually in a horse barn, where equipment and supplies are stored.

Tagging   Shearing the wool from around the tail and the udder of a ewe prior to lambing or from around the tail of a lamb.

Tags   Small bits of manure caked to the hair of cattle or the wool of sheep. Usually found in the area of the rear quarters and/or tail.

Tail ender   An animal in the poorest portion (end) of a group.

Tail head   That point near the base of the tail where the tail begins its downward descent.

Tank trucks   Insulated trucks in which milk is transported from a dairy farm to a processing plant or from one processing plant to another.

Tattoo (tattooing)   An indelible mark or scar made by tattooing. A method of punching small holes with a die in the form of numbers or letters through the skin, usually in the inside of the ear and then filling them with tattoo ink. Gives a permanent mark for identification.

Teaser   A male (bull or ram) made incapable, by vasectomy or by use of an apron to prevent copulation, of impregnating a female.

Teat cup   A cup-shaped part of a milking machine that is placed over the teat during the milking act.

Teat dilator   A medicated wax or steel-chrome plated object placed in the teat opening of the cow's udder. Used after an operation on the teat to prevent closing of the teat canal by adhesion of the healing surfaces.

Teat dip   A germicidal solution used to prevent the spread of mastitis in which the cow's teats are dipped at the conclusion of the milking act.

Tendon   A tough cord of dense, white fibrous connective tissue uniting a muscle with some other part and transmitting the force which the muscle exerts.

Term   The gestation period.

Terramycin   An antibiotic. Also called tetracycline.

Testes (testicles)   Paired organs of the male that produce both sperm cells and the male sex hormone.

Tetany   A condition in an animal in which there are localized, spasmodic muscular contractions.

Tether   To tie an animal with a rope or chain to allow grazing but prevent straying.

Throwback   An individual organism that manifests certain characters peculiar to a remote ancestor but which have been in abeyance during one or more of the intermediate generations.

Thurl   That portion of a cow's anatomy just posterior and somewhat ventral to the hip bones. It is part of the rump and is located at the widest portion of the sacrum.

Thyroprotein   A protein (usually produced by treating casein with iodine) that has thyroid activity. Thyroprotein stimulates metabolic processes in the body and is sometimes used to increase milk production in dairy cows.

Toe-out   To walk with the feet pointed outward.

Total digestible nutrients (TDN)   A term used in animal feeding that designates the sum of all the digestible organic nutrients. (Digestible fat is multiplied by 2.25 because of its higher energy content.) TDN is a way of expressing the energy content of a feed.

Toxicants   Compounds, usually man-made, found in the environment that are toxic to domestic farm animals.

Toxins   Poisons, products of cell metabolism, which are toxic to animals.

Trace elements   See *minor elements.*

Trichinosis   A disease of swine caused by the tissue invasion of an internal parasite, *Trichina spiralis.* The disease is uncommon, but meat from infected animals is unfit for human consumption unless thoroughly cooked.

True stomach (abomasum)   The fourth compartment of the stomach of ruminants. The abomasum secretes digestive enzymes comparable to those produced by the stomach of simple-stomached (monogastric) animals.

Tucked up   Contracted; appearing as if drawn in or up; said specifically of the flanks or walls of the abdomen of a cow or other animal.

Tusks   Extensions of the canine teeth in the male hog. These become particularly large in boars over 1 yr of age.

Twin band   During lambing, those ewes with twins that are often placed in a separate group because they need extra feed and care.

Twist   The inner bulge of the rear quarters of an animal. The inner side of the thighs.

Type (livestock) The conformation of an animal that indicates or suggests the purpose it serves. In animal husbandry, an ideal embodying all the characteristics that make an animal highly useful for a specific purpose.

Type classification A program usually sponsored by a breed association, whereby a registered animal's conformation is compared by an official inspector (classifier) with the "ideal" or "true" type animal of that breed. Most such programs are for dairy cattle.

U

Udder The anatomical structure of the female that contains the mammary gland.

Ultrasonic estimation of backfat A method for estimating backfat in live pigs. Relies on ultrasonic techniques rather than skin incision and manual measurements. Also used for estimating fatness in cattle.

United States Department of Agriculture (USDA) Department of the executive branch of the U.S. government. Deals with factors affecting production, processing, marketing, consumption, and utilization of food and fiber.

Unsoundness Defect or injury interfering with animal usefulness.

Unthriftiness Lack of vigor; poor growth or development.

USDA sire summaries Calculations of the genetic worth of dairy bulls prepared by the Animal Improvement Programs Laboratory, ARS, USDA. These summaries present numerical indexes used to guide breeding programs. The indexes are computed by analyzing the production records of the daughters of bulls entered in the National Cooperative Dairy Herd Improvement Program.

V

Vagina In female mammals, a canal leading from the uterus to the external orifice (vulva) of the genital canal.

Vas deferens In the male animal, the duct carrying the spermatozoa from the testicles to the urethra in the penis.

Veal A calf fed for early slaughter (about 3 mo).

Villi Numerous fingerlike projections extending from the intestinal wall into the lumen of the small intestine. Nutrients are absorbed from the intestine into the bloodstream through the villi.

Viral diseases Diseases, such as African swine fever, foot and mouth disease, and hog cholera, caused by viruses.

Volatile fatty acids Short chain organic acids that are volatile upon stream distillation. These compounds are formed in the rumen as a result of bacterial degradation of cellulose and other carbohydrates. Examples are acetic, propionic, and butyric acid.

Vulva The vestibule, or duct, in most female mammals, common to both genital and urinary tracts and leading to the body surface.

W

Wattles In swine, the small tissue projections hanging in pairs from the neck area. Presence of wattles is an inherited trait.

Weaning Taking the nursing young away from the mother and depriving it of the opportunity to nurse. This term is also used when calves are removed from diets containing fluid milk or milk replacer.

Weigh-a-Day-a-Month A production testing program used by dairy producers based on the recording of milk weights of each cow for one day each month. No milk analyses are made.

Wet ewe Ewe with suckling lamb; lactating ewe.

Wether A male sheep that was castrated at an early age, usually within 2 weeks after birth.

Wholesale cuts of meat Cuts of meat that are marketed at the wholesale level. Wholesale cuts are further subdivided for sale at the retail level.

Wing shoulder A condition in which the shoulder joint is away from the rib structure and skeleton.

Wisconsin Mastitis Test A semiquantitative test of cow's milk based on its leucocyte content and indicating the freedom from or presence of mastitis in the cow.

With calf Terminology designating a cow that is pregnant.

Wry tail Tailhead set either to the right or left of center.

X

X chromosome A sex chromosome found in mammals. All of the eggs produced by the female contain X chromosomes. When two X chromosomes are present in the fertilized egg, a female develops.

Y

Y chromosome A sex chromosome found only in male mammals. When a fertilized egg contains an X chromosome and a Y chromosome, a male develops. About 50% of the sperm produced by the male contain Y chromosomes.

Yolk The natural grease and suint covering of the wool fibers of the unscoured fleece and excreted from glands in the skin.

Young herd (swine shows) A show ring classification consisting of a boar and three sows, farrowed on or after 1 September of the year preceding the show.

Z

Zero grazing   A term applied to the practice of cutting or chopping green forage crops in the field and hauling this green feed to cows daily.

References

Bath, D. L., Dickinson, F. N., Tucker, H. A., and Appleman, R. D. (1978). "Dairy Cattle: Principles, Practices, Problems, Profits," 2nd ed. Lea and Febiger, Philadelphia.

Blakely, J., and Bade, D. H. (1982). "The Science of Animal Husbandry," 3rd ed. Reston Publishing Co., Virginia.

Bogart, R., and Taylor, R. E. (1983). "Scientific Farm Animal Production," 2nd ed. Burgess Publishing Co., Minneapolis.

Bundy, C. E., Diggins, R. V., and Christenson, V. W. (1982). "Livestock and Poultry Production," 5th ed. Prentice-Hall Inc., Englewood Cliffs, New Jersey.

Campbell, J. R., and Marshall, R. T. (1975). "The Science of Providing Milk for Man." McGraw-Hill, New York.

Cole, H. H., and Garrett, W. N. (1980). "Animal Agriculture: The Biology, Husbandry, and Use of Domestic Animals." H. W. Freeman and Company, San Francisco.

Ensminger, M. E. (1970). "Sheep and Wool Science," 4th ed. Interstate Printers and Publishers, Danville, Illinois.

Ensminger, M. E. (1971). "Dairy Cattle Science." Interstate Printers and Publishers, Danville, Illinois.

Ensminger, M. E. (1976). "Beef Cattle Science," 5th ed. Interstate Printers and Publishers, Danville, Illinois.

Ensminger, M. E. (1983a). "Animal Science," 8th ed. Interstate Printers and Publishers, Danville, Illinois.

Ensminger, M. E. (1983b). "The Stockman's Handbook," 6th ed. Interstate Printers and Publishers, Danville, Illinois.

Ensminger, M. E., and Parker, R. O. (1983). "Swine Science," 5th ed. Interstate Printers and Publishers, Danville, Illinois.

Frandsen, J. H. (1958). "Dairy Handbook and Dictionary." Nittany Printing and Publishing Company, State College, Pennsylvania.

Juergenson, E. M. (1981). "Approved Practices in Sheep Production," 4th ed. Interstate Printers and Publishers, Danville, Illinois.

Krider, J. L., Conrad, J. H., and Carroll, W. E. (1982). "Swine Production," 5th ed. McGraw-Hill, New York.

National Pork Producer's Council (1982). "Pork Facts, Production and History: Glossary of Production Terms." Des Moines, Iowa.

Neumann, A. L. (1977). "Beef Cattle," 7th ed John Wiley and Sons, New York.

O'Mary, C. C., and Dyer, I. A. (1978). "Commercial Beef Cattle Production," 2nd ed. Lea and Febiger, Philadelphia.

Pond, W. G., and Maner, J. K. (1984). "Swine Production and Nutrition." AVI Publishing Company, Westport, Connecticut.

Trimberger, G. W., and Etgen, W. M. (1983). "Dairy Cattle Judging Techniques," 3rd ed. Prentice-Hall Inc., Englewood Cliffs, New Jersey.

Yapp, W. W. (1959). "Dairy Cattle Judging and Selection." John Wiley and Sons, Inc., New York.

Breeds and Genetics

Animal Breeds

Paul A. Putnam

U.S. Department of Agriculture
Agricultural Research Service
Stoneville, Mississippi

Introduction

Tables 1–7 of breeds of livestock have been excerpted and abstracted from a much more complete listing by I. L. Mason in *A World Dictionary of Livestock Breeds: Types and Varieties.* You are referred to Mason's text if you are interested in more detail or the less-recognized breeds. Additional information on breeds of livestock, especially cattle, may be obtained from the other references cited.

Handbook of Animal Science

Table 1

Ass (*Equus asinus*)

Breed Name	Origin	Primary Use	Identifying Characteristics	Breed Society or Herd Book Established
American	United States	—	—	1888
Andalusian	Spain	—	grey	—
Brazilian	Brazil	—	—	—
Catalan	Spain	—	black, dark grey, or brown with pale underparts	1880
Cyprus	Africa	—	—	—
Hamadan	Iran	—	usually white	—
Leon-Zamora	Spain	—	black with paler muzzle and underparts; long hair	1941
Martina Franca	Italy	—	nearly black with light underparts	1943
Pega	Brazil	—	roan or dark grey	1949
Poitou	France	—	black or dark brown; long hair	—
Sicilian	Italy	—	nearly black with pale muzzle and belly, some-times grey	—

Table 2

Buffalo (*Bubalus bubalus*)

Breed Name	Origin	Primary Use	Identifying Characteristics	Breed Society or Herd Book Established
Egyptian	Egypt	d.dr.	grey-black; short, curved horns	—
European	Europe	d.m.dr.	dark grey to black, often white on head, tail, and feet	—
Jafarabadi	India	d.	usually black; large, drooping horns	—
Kundi	Pakistan	d.	usually black, occasionally brown, often with white on head, tail, and feet	—
Mehsana	India	d.	black, occasionally brown or grey	—
Murrah	India	d.	black, short, coiled horns	1940
Nagpuri	India	dr.d.	usually black, sometimes white on face, legs, and tail; long horns	—
Nili-Ravi	Pakistan, India, Punjab	d.	usually black, occasionally brown, often with walleyes and white marks on head, legs, and tail	—
Surti	India	d.	black or brown, usually with two white collars	—
Tarai	India, Nepal	d.	black, occasionally brown, with white tail	—

Table 3
Cattle (*Bos taurus*)[a]

Breed Name	Origin	Primary Use	Identifying Characteristics	Breed Society or Herd Book Established
Aberdeen-Angus	Scotland	m.	black; polled	1862
Africander	Africa	m.dr.	usually red; long, lateral horns	1907
American Brown Swiss	United States	d.	—	1880
Angeln	Germany	d.	red	1879
Angoni[b]	L. Malawi	m.d.dr.	many colors	—
Ankole[b]	Africa	—	often red, also fawn, black, or pied	—
Aosta	Italy	d.m.	black pied, red pied	1937
Apulian	Italy	dr.m.	grey	1931
Aquitaine Blond	France	m.dr.	yellow/yellow brown	—
Asturian	Spain	dr.d.m.	shades of red with paler extremities	1933
Aubrac	France	dr.d.m.	fawn to brown	1914
Aulie-Ata	USSR	d.	usually black pied, occasionally red pied or black	1935
Australian Illawarra Shorthorn	New South Wales	d.	usually red or roan	—
Ayrshire	Scotland	d.	red-brown and white; lyre horns	1877
Bachaur	India	dr.	grey-white	—
Balinese[c]	Indonesia	m.dr.	red turning black	—
Bargur	India	dr.	usually red-and-white spotted	—
Beef Shorthorn	Scotland	m.	—	1882
Belted Galloway	Scotland, England	—	black or dun with dominant white belt	1922
Bestuzhev	USSR	m.d.	red, often with marks on head, belly, and feet	1928
Blacksided Trondheim and Nordland	Norway	d.	white with black (occasionally red) side or spots on side; polled	1943

Breed	Country/Region	Type	Color	Year
Bonsmara[d]	Africa	m.	red	—
Boran[b]	Kenya, Ethiopia, Somalia	m.	usually white or grey, also red or pied	—
Brahman[b]	United States	—	usually silver-grey	1924
Brazilian Polled[b]	Brazil	m.d.	yellow	1939
Breton Black Pied	France	d.	black pied	1919
British Friesian	Great Britain	d.m.	black and white	1909
British White	Great Britain	d.m.	white with black points	1918
Campine Red Pied	Belgium	d.m.	red pied; some Shorthorn blood	1919
Caracu	Brazil	m.d.dr.	usually yellow; long horns	1916
Central and Upper Belgium	Belgium	d.m.	white, blue pied, or blue	1919
Charolais	France	m.dr.d.	white (or cream) with pink mucosa	1864
Chiana	Italy	m.dr.	white with black points and mucosa	1956
Dairy Shorthorn	Great Britain	d.m.	variant of Shorthorn	1905
Damascus	Syria	d.	brown	—
Danish Black Pied	Denmark	d.	black pied	1902
Danish Red	Denmark	d.	red	1885
Danish Red Pied	Denmark	d.	red pied	—
Devon	England	m.	cherry red	1851
Dexter	Ireland	d.m.	red or black; short-legged	1887
Dhanni[b]	Pakistan	dr.	white with black markings, usually spotted or color-sided, black with white markings, red-and-white, or black	—
Drakensberger[d]	Africa	m.d.	black	1948
Droughtmaster[d]	Australia	m.	red; horned or polled	1956
Dutch Black Pied	Netherlands	d.m.	black pied	1874
East Anatolian Red	Turkey	d.m.	—	—
East Flemish Red Pied	Belgium	d.m.	red pied	1900
Eastern Spotted	France	d.m.dr.	variant of Eastern Red Pied	1930
Egyptian	Egypt	dr.d.	yellow-brown to black	—
Finnish	Finland	d.	red-and-white, white, red	1946

(continued)

111

Table 3 (*Continued*)

Breed Name	Origin	Primary Use	Identifying Characteristics	Breed Society or Herd Book Established
Finnish Ayrshire	Finland	d.	variant of Ayrshire	1901
Flemish	France	d.	mahogany	1886
French Brown	France	d.m.dr.	brown	1911
French Friesian	France	d.	black and white	1922
Fribourg	Switzerland	d.m.dr.	black pied	—
Galician Blond	Spain	dr.m.d.	cream to golden red	1933
Galloway	Scotland	m.	black or brownish black; polled	1862
Garfagnana	Italy	d.m.dr.	grey	1935
Gascony	France	dr.m.	grey; with or without black points and mucosa	1894
German Black Pied	Germany	d.m.	black and white lowland and pied	1878
German Red Pied	Germany	d.m.	red pied	1922
Gir[b]	India	d.dr.m.	mottled red and white	—
Gorbatov Red	USSR	d.m.	red	1921
Grey Adige	Italy	dr.d.m.	grey	1934
Groningen Whiteheaded	Netherlands	d.m.	black (5% red); white head and belly	1906
Guernsey	Channel Islands	d.	brown and white	1842
Haryana[b]	India	dr.d.	grey-white short-horned	—
Hereford	England	m.	red; white head and belly	1846
Herens	Switzerland	d.m.dr.	dark red-brown	—
Highland	Scotland	m.	usually red-brown to brindle dun or black; long horns and hair	1884
Hinterwald	Germany	d.dr.	red or yellow mottled; head and legs white	1901
Holando-Argentino	Argentina	d.m.	variant of Friesian	—
Holstein-Friesian	United States, Canada	d.	black and white	1871
Hungarian Pied	Hungary	d.m.dr.	red and white	1896

112

Breed	Country	Use	Description	Year
Hungarian Simmental	Hungary	d.m.	—	1920
Icelandic	Iceland	d.	red or red-and-white, also brindle, brown, black, grey, white, or pied; usually polled	1903
Indo-Brazilian[b]	Brazil	m.	white to dark grey	1938
Israeli Friesian	Israel	d.	—	1951
Italian Brown	Italy	d.m.	variant of Swiss Brown	1956
Italian Friesian	Italy	d.m.	black pied; variant of Friesian	1956
Jamaica Black[d]	West Indies	m.	polled	1954
Jamaica Brahman[b]	West Indies	m.	—	1949
Jamaica Hope[d]	West Indies	d.	—	1953
Jamaica Red[d]	West Indies	m.	polled	1952
Japanese	Japan	m.dr.d.	black, brown; polled	—
Jersey	Channel Islands	d.	fawn, mulberry, or grey; often black skin pigment	1844
Kangayam[b]	India	dr.	white or grey with red calves	—
Kankrej[b]	India	dr.d.	grey	—
Kenana[b]	Sudan	d.dr.	light blue-grey with black points	—
Kenya Boran[b]	Africa	m.	improved variant of Boran	1951
Kerry	Ireland	d.	black	1887
Kholmogor	USSR	d.	usually black pied, also red pied, black, or white	—
Krishna Valley[b]	India	dr.d.	grey-white short-horned	—
Kuri	Africa	m.d.	light or white, sometimes with spots; gigantic, bulbous horns	—
Latvian Brown	USSR	d. or d.m.	brown or dark red	1911
Limousin	France	m.dr.	dark yellow-brown	1886
Lincoln Red	England	m.	red	1822
Madurese[e]	Indonesia	dr.m.	chestnut with pale underparts; small hump	—
Maine-Anjou	France	m.	—	—
Marche	Italy	m.dr.	grey	1957
Maremma	Italy	dr.m.	grey; long, open lyre horns	1935
Mashona[b]	Rhodesia	m.	usually black or red; often polled	1950

(continued)

113

Table 3 (*Continued*)

Breed Name	Origin	Primary Use	Identifying Characteristics	Breed Society or Herd Book Established
Maure[b]	Senegal	pa.dr.d.	red, black, or pied; short-horned	—
Mauritius Creole	Mauritius	d.	usually white with black points or with colored flecks; also brown or brown-and-white; polled	—
Meuse-Rhine-Yssel	Netherlands	d.m.	red pied, red-and-white	1906
Middle German Red	Germany	d.m.dr.	red	1911
Modena	Italy	d.m.dr.	white with dark points	1957
Modica	Italy	dr.m.d.	—	1952
Modica-Sardinian	Sardinia	dr.m.d.	yellow-brown to dark red	1936
Mongolian	China	d.m.dr.	usually brindle or reddish-brown, somtimes black, yellow, or pied	—
Murcian	Spain	m.dr.	dark red	—
Murnau-Werdenfels	Germany	d.m.dr.	red-yellow; local breed similar to German Brown but smaller and more red-yellow in color	—
Murray Grey	Australia	m.	silver-grey to grey-dun; polled	1962
Nagori[b]	India	dr.	grey-white; short-horned	—
N'Dama[b]	Africa	m.dr.	usually fawn, also light red or dun; larger (plains) variant with longer lyre-shaped horns; smaller (hill) variant with shorter lyre or crescent horns	—
Nguni[b]	Natal, Rhodesia, Africa	d.dr.	often color-sided	—
Normandy	France	d.m.	red-brown and white; often brindled; usually red spectacles	1883
Northern Dairy Shorthorn	England	d.m.	—	1944

114

Breed	Country	Use	Description	Year
Norwegian Red	Norway	d.	red (or red with some white); horned or polled	—
Oberinntal Grey	Austria	d.m.dr.	light grey	1924
Ongole[b]	India	dr.d.	grey-white; short-horned	—
Ovambo[b]	Africa	—	usually dun or black	—
Parthenay	France	d.m.dr.	fawn	1893
Piedmont	Italy	m.d.dr.	white or pale grey with black points	1887
Pinzgau	Austria, Germany, Italy	d.m.dr.	red-brown; color-sided with colored head	1963
Polish Black-and-White Lowland	Poland	d.m.	black pied	—
Polish Red	Poland	d.m.dr.	red	—
Polish Red-and-White Lowland	Poland	d.m.dr.		—
Pontremoli	Italy	dr.m.d.	yellowish-brown with pale black stripe	1935
Ponwar[b]	India	dr.	black and white	—
Pyenean	Spain	m.dr.d.	red to cream; pale extremities and under-parts	1933
Rath[b]	India	dr.d.	grey-white	—
Red Pied Friuli	Italy	m.d.dr.	red pied	1931
Red Poll	England	d.m.	red	1873
Red Sindhi[b]	Pakistan	d.		—
Red Steppe	USSR	d.	red	1923
Reggio	Italy	d.dr.m.	red to yellow-brown	1935
Romagna	Italy	m.dr.	grey to white	1956
Rouge de l'Quest	France	m.d.	red	1966
Russian Black Pied	USSR	d.	black pied	1940
Sahiwal[b]	Pakistan, Senegal, India, Philippines	d.	usually reddish-dun, usually with white markings	—
Salers	France	d.dr.m.	mahogany	1908
San Martin	Colombia	m.d.	yellow-red or chestnut	—
Santa Gertrudis[d]	United States	m.	red	1951

(continued)

Table 3 (*Continued*)

Breed Name	Origin	Primary Use	Identifying Characteristics	Breed Society or Herd Book Established
Senepol[d]	St. Croix (Virgin Islands)	m.	red	1945
Shetland	Scotland	d.m.	various colors	1911
Shorthorn	England	m.d.	red, roan, white, sometimes red-and-white or roan-and-white	1822
Shuwa[b]	Nigeria	pa.m.d.	dark red	—
Simmental	Switzerland	d.m.dr.	dun red or leather-yellow and white	1890
Sinhala[b]	Ceylon	—	usually black or red, sometimes pied; dwarf	—
Sokota[b]	Nigeria, Ghana	d.dr.	grey or dun	—
South Anatolian Red	Turkey	d.	—	1891
South Devon	England	d.m.	red	1941
Suksun	USSR	d.	red	1879
Sussex	England	m.dr.	red	1891
Swedish Friesian	Sweden	d.m.	black and white	1949
Swedish Jersey	Sweden	d.	—	1938
Swedish Polled	Sweden	d.	white, red	1928
Swedish Red-and-White	Sweden	d.	red with small white markings	1878
Swiss Brown	Switzerland	d.m.dr.	grey-brown	1931
Tagil	USSR	d.	usually black pied, also black, red, or red pied	

116

Tarenntaise	France	d.m.	fawn to yellow	1880
Telemark	Norway	d.	color-sided usually red or yellow; lyre horns	1926
Tharparkar	Pakistan	d.dr.	grey, lyre-horned	—
Tonga[b]	Zambia	—	usually red, black, or pied	—
Tudanca	Spain	dr.m.d.	yellow, fawn or chestnut; silver spots; dark extremities	—
Tuli[b]	Rhodesia	—	yellow, golden-brown, or red	—
Turino	Portugal, Brazil	d.m.	—	—
Ukrainian Grey	USSR	m.dr.d.	grey	1935
Ukrainian Whiteheaded	USSR	d.	red (or black) with white head, feet, and belly	1926
Vorderwald	Germany	d.dr.	red to black; blaze on face	—
Welsh Black	Wales	m.d.	—	1874
West Flemish Red	Belgium	d.m.	—	1919
White Fulani[b]	Ghana, Nigeria	dr.d.m.	usually white with black skin and points	—
Yaroslavl	USSR	d.	usually black (occasionally red) with white head and feet	1925
Yugoslav Pied	Yugoslavia	—	pied, red, and white	—
Yurino	USSR	d.m.	red or brown	1937

[a]Unless otherwise noted. [b]*Bos indicus*. [c]*Bos javanicus*, a domesticated Banteng. [d]Based on *Bos taurus* × *Bos indicus* foundation. [e]Based on *Bos javanicus* × *Bos indicus* foundation.

117

Table 4
Goat (*Capra hircus*)

Breed Name	Origin	Primary Use	Identifying Characteristics	Breed Society or Herd Book Established
Agrigento	Italy	d.m.	rufous and white, or off-white with brown spots; screw horns	—
Anatolian Black	Turkey	m.fl.	usually black, also brown, grey, or pied	—
Anglo-Nubian	Great Britain	d.		1890
Angora	Turkey	fl.m.	—	1900
Appenzell	Switzerland	d.	white; long hair; polled	—
Apulian	Italy	d.	usually pied; horned or polled; short hair; brown with V-shaped screw horns or short, backward-curving horns	—
Assam Hill	India	m.	usually white; long hair; short ears and horns	—
Baluchi	Pakistan	fl.d.	usually black, also brown, grey, or white	—
Barbari	India, Pakistan	d.	often white with red spots; short coat, ears, and horns	—
Bari	Pakistan	d.m.	usually white, sometimes pied; short ears and coat	—
Beetal	Pakistan	d.m.	usually red, tan, or pied; long ears	—
Belgian Fawn	—	d.m.	polled	1931
Bengal	Pakistan, India	m.	usually black, also brown, white, or fawn; short coat and ears	—
British Alpine	Great Britain	d.	black with light points and face stripes; horned or polled	1925
British Saanen	Great Britain	d.	white; horned or polled	1925
British Toggenburg	Great Britain	d.	brown with light points and face stripes; polled or horned	1925
Campine	Belgium	—	polled	1931

118

Name	Country	Use	Description	Year
Chamois Coloured	Switzerland	d.m.	brown with black face stripes, back stripe, belly, and legs; horned or polled	—
Chaper	Pakistan	m.	usually black	—
Cheghu	India	fl.m.dr.	usually white	—
Damani	Pakistan	d.	usually black	—
Damascus	Syria, Lebanon	d.	usually red or brown, also pied or grey; usually polled	—
Dera Din Panah	Pakistan	d.fl.m.	usually black, also red; long ears	—
Dole	Norway	d.	usually blue pied or brown; polled	—
Don	USSR	fl.d.	usually black, also pied, grey, or white	1934
Dutch Toggenburg	Netherlands	d.	—	—
Dutch White	Netherlands	d.	polled	—
French Alpine	France	d.	white, fawn, or pied; horned or polled	—
Gaddi	India	fl.m.	usually white, also grey or red; lop ears	—
German Improved Fawn	Germany	d.	red-brown to fawn with black face stripe, belly, and feet; polled	—
German Improved White	Germany	d.	polled	—
Granada	Spain	d.	black; polled	—
Grisons Striped	Switzerland	d.m.	black with pale face stripe and legs; horned or polled	—
Gujarati	India	d.m.fl.	long, black hair; white lop ears; corkscrew horns; Roman nose	—
Improved North Russian	USSR	d.	white	—
Jamnapari	India	d.m.	long lop ears	—
Kaghani	Pakistan	fl.	white, grey, brown, or black	—
Kamori	Pakistan	d.m.fl.	usually grey-and-brown, also black-and-white; long ears	—
Kashmiri	India	fl.	white or black-and-white	—
Kirgiz	USSR	fl.m.d.	—	—
Leri	Pakistan	m.d.fl.	usually black; long ears	—
Malaga	Spain	d.	red; prisca horns	—
Mingrelian	USSR	d.	usually white, roan, or grey	—

(continued)

119

Table 4 (*Continued*)

Breed Name	Origin	Primary Use	Identifying Characteristics	Breed Society or Herd Book Established
Murcian	Spain	d.	usually black and polled	—
North Russian	USSR	d.	usually white	—
Osmanabad	India	m.d.	usually black, often pied; long- and short-haired	—
Poitou	France	d.	black-brown with pale legs; long hair; polled	—
Red Bosnian	Yugolsavia	d.	red, grey, black, brown, or pied	—
Red Sokoto	Nigeria	—	—	—
Saanen	Switzerland	d.	white; polled	—
Salt Range	Pakistan	d.fl.	black-and-white or black; log ears; vertical screw horns	—
Sirli	Pakistan	fl.m.	lop ears	—
Sirohi	India	d.	white or brown; lop ears	—
Somali	Somalia, Ogadan, Kenya	m.	white, occasionally with colored spots or patches; short hair and ears; males horned, females horned or polled	—
Soviet Mohair	USSR	fl.m.	white	—
Surti	India	d.	short hair and ears	—
Syrian Mountain	Syria, Lebanon, Anatolia, Israel, Jordan, Iraq	m.d.fl.	—	—
Telemark	Norway	d.	white; polled	—
Thori	Pakistan	m.d.	polled, usually red	—
Toggenburg	Switzerland	d.	brown to mouse-grey with white face stripes and feet; polled	—
Valais Blackneck	Switzerland	m.d.	front half black, hind part white; long hair	—
Verzasca	Switzerland	d.m.	black	—
Zaraibi	Egypt	d.	polled	—

120

Table 5
Horse (*Equus caballus*)

Breed Name	Origin	Primary Use	Identifying Characteristics	Breed Society or Herd Book Established
Akhal-Teke	USSR	l.	—	—
Albanian	—	py.	—	—
Altai	USSR	dr.d.m.	—	—
American Quarter Horse	United States	l.	—	1940
American Saddle Horse	United States	l.	bay, brown, chestnut, grey, or black	1891
American Trotter/ Standardbred	United States	l.	—	1871
Anatolian Native	Turkey	py.	—	—
Andalusian	Spain	l.	usually brown, grey, or black	—
Appaloosa	United States	l.	color type; usually chubary (white rump usually with small spots), also white with spots over whole body, or colored with spots over body or rump only, or blue roan	1938
Arab	—	l.ri.	—	1919
Argentine Criollo	Argentina, Uruguay	l.	—	—
Auxois	France	h.dr.	bay, roan, or grey	1913
Azerbaijan	USSR	l.py.	grey	—
Baluchi	Pakistan	l.	turned-in ears	—
Barb	Africa	l.	—	—
Bashkir	USSR	d.py.	—	—
Basque-Navarre Pony	Spain	—	usually dark bay or black	—
Belgian	Belgium	h.dr.	—	1886
Bhotia Pony	Nepal	ri.pa.	similar to Tibetan Pony but less broad; often white or grey	—

(*continued*)

121

Table 5 (*Continued*)

Breed Name	Origin	Primary Use	Identifying Characteristics	Breed Society or Herd Book Established
Black Sea	USSR	l.	—	—
Bosnian Pony	Yugoslavia	—	brown or dun, white, black, or chestnut	—
Boulonnais	France	h.dr.	usually grey	1886
Breton	France	h.–l.dr.	—	—
Budennyi	USSR	l.	usually chestnut or bay	—
Burguete	Spain	l.dr.	usually black or bay	—
Campolino	Brazil	l.dr.	usually bay, sorrel, or chestnut	—
Canadian	Canada	l.dr.	—	—
Chumysh	USSR	l.dr.	—	—
Cleveland Bay	England	l.dr.	—	1884
Clydesdale	Scotland	h.dr.	usually bay, brown, or black, with white on face and feet	1877
Comtois	France	h.–l.dr.	usually bay	—
Connemara Pony	Ireland	—	usually grey, black, bay, brown, or dun	1923
Crioulo	Brazil	py.	—	—
Cukurova	Turkey	l.	—	—
Dales Pony	England	—	usually black, dark brown, grey, or dark bay	1916
Dartmoor Pony	England	—	usually black, brown, bay or grey	1899
Dole	Norway	l.dr.	usually brown or bay, also black, occasionally chestnut	1902
Don	USSR	l.	—	—
Dongola	Sudan, Eritrea	l.	reddish-bay, often with white face-blaze and feet	—
Dutch Draft	Netherlands	h.dr.	—	1915
East Bulgarian	Bulgaria	l.ri.dr.	—	—
East Friesian	Germany	l.dr.	—	—
Estonian Pony	Estonia	—	—	—

Breed	Origin	Type	Color	Year
Exmoor Pony	England	—	bay, dun, or brown; mealy nose; no white markings	1921
Fell Pony	England	dr.	—	1893
Finnish	Finland		usually chestnut, sometimes brown	—
Fjord	Norway	py.	usually dun with dark mane, tail, and back stripe	1910
Frederiksborg	Denmark	l.dr.	usually chestnut	—
Freiberg	Switzerland	l.dr.		—
French Ardennes	France	h.dr.		1908
French Saddle Horse	France	l.ri.		—
French Trotter	Normandy	l.		1922
Friesian	Netherlands	l.dr.ri.	black	1879
Galician Pony	Spain	—		—
Gelderland	Netherlands	l.ri.dr.	usually chestnut or grey, often white on legs and face	1880
Gotland Pony	Sweden	—	black, brown, or bay	1910
Groningen	Netherlands	l.ri.	usually black or brown	1880
Hackney	England	l.dr.		1883
Hafling	Austria, Italy, Germany	py.	chestnut with light mane and tail	—
Hanover	Germany	l.		1888
Highland Pony	Scotland	—	usually grey, dun, black, or brown, with eel stripe	1889
Hirzai	Pakistan	l.	white or grey	—
Holstein	Germany	l.		1886
Hungarian Draft	Hungary	h.dr.		—
Hutsul	Carpathians	py.		—
Iceland Pony	—		chestnut, also brown, black, grey, or dun	—
Iomut	USSR	l.		—
Jutland	Denmark	h.dr.	usually chestnut	1888
Kabarda	USSR	l.		—
Karabair	USSR	l.		—

(continued)

123

Table 5 (*Continued*)

Breed Name	Origin	Primary Use	Identifying Characteristics	Breed Society or Herd Book Established
Karabakh	USSR	l.	—	—
Karelian	USSR	—	—	—
Kathiaware	India	l.	usually dun; turned-in ears	—
Kazakh	USSR	py.	—	—
Kirgiz	USSR	py.	—	—
Kushum	USSR	l.ri.dr.m.	—	—
Kustanai	USSR	l.	—	—
Kuznetsk	USSR	l.	—	—
Latvian Coach	USSR	l.dr.	—	—
Lipitsa	Europe	l.	usually grey	—
Lithuanian Draft	Lithuania, USSR	dr.	—	—
Lokai	USSR	l.	—	—
Losa	Spain	py.	—	—
Malopolski	Poland	l.dr.	—	—
Mangalarga	Brazil	l.	—	1934
Marwari	India	l.	usually chestnut or bay, also grey or brown, occasionally pied or cream	—
Mezen	USSR	—	—	—
Mingrelian	USSR	py.	—	—
Minusinsk	USSR	l.	—	—
Mongolian Pony	China	ri.d.dr.pa.	all colors	—
Morgan	United States	l.	—	1909
New Forest Pony	England	—	—	1891
Nonius	Europe	l.dr.	brown, dark brown, or black	—
Noric	Austria	h.dr.	usually brown, bay, or chestnut, also black, grey, or tiger-spotted	—
Norman Cob	France	l.dr.	—	—

Breed	Country	Use	Description	Year
North Swedish	Norway	l.dr.	—	1909
Ob	USSR	—	—	—
Oldenburg	Germany	l.dr.	—	—
Orlov-Rostopchin	USSR	l.	—	—
Orlov-Trotter	USSR	l.	—	—
Pechora	USSR	—	—	—
Percheron	France	h.dr.	usually grey, white, or black	1833
Poitou	France	h.dr.	—	1885
Polesian	USSR	py.	—	—
Portuguese	Portugal	l.ri.dr.	—	—
Rhenish	Germany	h.dr.	—	1876
Romanian	Romania	l. or py.	—	—
Rottal	Germany	l.dr.	—	—
Russian Draft	USSR	—	—	—
Russian Trotter	USSR	l.	—	—
Sandalwood Pony	Indonesia	—	—	—
Sardinian	—	l.	usually bay, chestnut, or grey	—
Schleswig	Germany	h.dr.	usually chestnut	—
Shan Pony	Burma	—	—	—
Shetland Pony	Scotland	—	dwarf	1890
Shire	England	h.dr.	black, brown, bay, or grey	1878
Sicilian	Italy	l.	—	—
South China Pony	China	pa.dr.	—	—
South German Coldblood	Germany	h.–l.dr.	—	—
Soviet Draft	USSR	h.	—	—
Spiti Pony	India	—	usually grey or dun, occasionally bay or black	—
Suffolk	England	h.dr.	chestnut	1877
Swedish Ardennes	—	h.dr.	—	1901
Tavda	USSR	—	—	—
Tennessee Walking Horse	United States	l.	—	1938
Tersk	USSR	l.	silver-grey	—
Thoroughbred	England	l.ri.	—	1791

(continued)

125

Table 5 (*Continued*)

Breed Name	Origin	Primary Use	Identifying Characteristics	Breed Society or Herd Book Established
Tori	USSR	l.dr.	—	—
Trakenhnen	Germany	—	—	—
Transbaikal	USSR	—	—	—
Tushin	USSR	l.–py.	—	—
Vladimir Draft	USSR	h.	—	—
Voronezh Coach	USSR	h.–l.dr.	—	—
Vyatka	USSR	—	—	—
Waziri	Pakistan	l.	—	—
Welsh Pony	Wales	ri.	—	1902
Western Sudan Pony	Sudan	l.–py.	usually light bay, chestnut, or grey, with white markings	—
White-Russian Coach	USSR	l.dr.	—	—
Wielkopolski	Poland	l.dr.	—	—
Wurttemberg	Germany	l.dr.ri.	usually brown, bay, chestnut, or black, with white markings	1895
Yakut	USSR	—	—	—
Zemaitukai	USSR	py.	—	—
Zweibrucken	Germany	l.	—	—

Table 6
Swine (*Sus scrofa*)

Breed Name	Origin	Primary Use	Identifying Characteristics	Breed Society or Herd Book Established
Alentejo	Portugal	—	red, golden, or black	—
American Landrace	United States	m.b.	—	1950
Andalusian Blond	Spain	—	whitish to golden	—
Andalusian Spotted	Spain	—	whitish with irregular black speckles, especially on head and rump	—
Angeln Saddleback	Germany	—	black head and rump	1928
Asturian	Spain	m.	black	—
Belgian Landrace	Belgium	b.	—	—
Beltsville No. 1	United States	m.	black with white spots; moderate lop ears	1951
Beltsville No. 2	United States	m.	red with white underline, occasionally black spots; short, erect ears	1952
Berkshire	England	m.p.b.	black with white points	1884
Bisaro	Portugal	—	black, white, or pied	—
Black Iberian	Spain	ld.m.	—	—
Breitov	USSR	p.ld.	lop ears	1953
British Landrace	Great Britain	—	—	—
British Saddleback	England	—	—	—
Bulgarian White	Bulgaria	m.	—	—
Canastra	Brazil	ld.	black	—
Canastrao	Brazil	—	often curly coated	—
Caserta	Italy	p.	black or grey; usually tassels	—
Chester White	United States	m.	—	1884
Danish Landrace	Denmark	b.	white, lop ears	—

(*continued*)

Table 6 *(Continued)*

Breed Name	Origin	Primary Use	Identifying Characteristics	Breed Society or Herd Book Established
Dermantsi Pied	Bulgaria	ld.	black with white points or white with black spots	—
Duroc	United States	m.	red	1883
Dutch Landrace	Netherlands	b.	—	1933
Dutch Yorkshire	Netherlands	p.	—	1913
East Balkan	Bulgaria	—	usually black, occasionally pied; prick ears	—
Estonian Bacon	USSR	b.	—	—
Extremadura Red	Spain	m.	—	—
Finnish Landrace	Finland	—	native lop-eared, may have blue spots	—
French Danish	France	b.	—	—
Galician	Spain	m.	small reddish or black spots on head and rump	—
German Landrace	—	b.p.	—	—
German Pasture	—	ld.	black head and rump; rough hair; prick ears	1899
Germany Yorkshire		p.b.	—	1914
Gloucestershire Old Spots	England	p.b.	—	1893
Hampshire	United States	m.	black with white belt	1934
Hereford	United States	ld.p.	red with white head, legs, belly, and tail	1923
Hungarian White	Hungary	p.b.	—	—
Kalikin	USSR	ld.	grey pied	—
Kemerovo	USSR	ld.	black or black pied	—
Lacombe	Canada	m.b.	lop ears	1959
Large Black	England	b.	black; lop-eared	1899
Large White	England	b.	prick ears	1884
Livny	USSR	ld.	occasionally black pied or grey-black; lop ears	—

128

Breed	Country	Type	Characteristics	Year
Majorcan	Spain	ld.	slate grey or black; tassels	—
Mangalitsa	Yugoslavia, Hungary, Romania	ld.	curly hair; lop ears	1927
Middle White	England	p.	—	1884
Minnesota No. 1	United States	m.b.	red	1946
Minnesota No. 2	United States	m.b.	black spotted	1948
Mirgorod	USSR	ld.	brown spotted	—
Montana No. 1	United States	m.	black	1948
National Long White Lop-eared	England	b.p.	white	1921
Nilo	Brazil	ld.	black; hairless	—
North Caucasus	USSR	ld.	black or spotted	—
North Siberian	USSR	ld.		—
Norwegian Landrace	—	b.	—	1956
Palouse	United States	b.		—
Pereira	Brazil	ld.	grey (or black), sometimes with red spots	—
Piau	Brazil	p.ld.	white (or grey or sandy) with black spots	—
Pietrain	Belgium	m.p.	dirty white with black or reddish spots, often red hairs; semi-lop ears	1950
Pirapitinga	Brazil	ld.	black or violet	—
Poland China	United States	m.	black with white points	1878
Polish Large White	Poland	b. or b.ld.		1956
Polish White Lop-eared	Poland	b.	—	—
Prestice	Czechoslovakia	m.ld.	black pied	—
Pulawy	Poland	ld.	black-and-white or black-and-red-and-white	—
Romagna	Italy	p.	dark brown or copper-colored	—
Russian Large White	USSR	m.		—
Russian Long-eared White	USSR	—	—	—
Russian Short-eared White	USSR	—	—	—

(continued)

Table 6 *(Continued)*

Breed Name	Origin	Primary Use	Identifying Characteristics	Breed Society or Herd Book Established
Siena Belted	Italy	—	black with white belt	—
South African Landrace	Africa	b.	—	1959
Spot	United States	m.	black with white spots	1914
Swabian-Hall	Germany	—	black with white belt	1925
Swedish Landrace	—	b.	—	1907
Swiss Improved Landrace	Switzerland	b.	—	—
Swiss Yorkshire	Switzerland	b.	—	—
Tamworth	England	m.b.	golden-red	1906
Taoyuan	Taiwan	ld.	black or dark grey; lop ears; wrinkled skin; straight tail	—
	China			
Tatu	Brazil	ld.	usually black and hairless	—
Turopolje	Croatia, Yugoslavia	ld.	white with black spots	—
Ukrainian Spotted Steppe	USSR	p.ld.	spotted and with bristles	—
Ukrainian White Steppe	USSR	p.ld.	—	—
Urzbum	USSR	—	white; lop-eared	—
Vitoria	Spain	m.	hairless	—
Welsh	Wales	b.	lop ears	1918
West French White	France	p.	—	1958
White-Russian Black Pied	USSR	m.ld.	—	—

Table 7
Sheep (*Ovis aries*)

Breed Name	Origin	Primary Use	Identifying Characteristics	Breed Society or Herd Book Established
Abyssinian	Ethiopia	m.d.	fat-tailed; usually brown; with mane; horned or polled	—
Algarve Churro	Portugal	m.cw.	white with black spots on face and feet, or black; horned	—
Algerian Arab	Algeria	m.cw.	♂ horned, ♀ polled	—
Altai	USSR	fw.m.	—	—
Altamura	Italy	d.m.cw.	occasionally dark spots on face; ♂ polled, or horned, ♀ polled	—
Amasya Herik	Turkey	cw.m.d.	usually white with dark spots on head	—
American Merino	United States	fw.	—	1906
American Rambouillet	United States	fw.m.	♂ horned, or polled, ♀ polled	1889
American Tuhnis	United States	m.	colored face and legs; polled	1896
Apulian Merino	Italy	fw.–(mw.)m.d.	♂ horned, ♀ polled	1942
Arabi	Iran	m.cw.	fat-tailed, black, pied, or white with black head; ♂ horned	—
Aragon	Spain	m.mw.	♀ polled	—
Argentine Merino	—	fw.m.	—	—
Arles Merino	France	fw.m.d.	usually horned, occasionally polled	—
Askanian	USSR	fw.m.	—	—
Aure-Campan	France	m.mw.	pale brownish-grey; occasionally black; usually horned or polled	—
Ausimi	Egypt	cw.m.	white with brown head, often brown neck, and occasionally brown spots; ♂ horned; ♀ polled; fat-tailed	—
Australian Merino	Australia, New Zealand	fw.	♂ horned or ♀ polled	1923

(*continued*)

Table 7 (*Continued*)

Breed Name	Origin	Primary Use	Identifying Characteristics	Breed Society or Herd Book Established
Avranchin	France	m.lw.	brownish face and feet; polled	1928
Awassi	Syria	m.d.cw.	fat-tailed; white with brown head and legs; sometimes black, white, grey or spotted face, occasionally all brown or black; horned; usually polled	—
Azerbaijan Mountain Merino	USSR	fw.	—	—
Azov Tsigai	USSR	mw.m.	—	—
Badano	Portugal	m.cw.d.	dirty white often with brown face and feet, occasionally black or brown; ♂ horned, ♀ polled	—
Balbas	USSR	m.d.cw.	fat-tailed; black or brown spots on face and feet; ♂ horned, ♀ polled	—
Balkhi	Pakistan	cw.d.	usually black or grey; horned (short)	—
Baluchi	Pakistan, Afghanistan	m.d.cw.	black marks on head and legs; ♂ horned, ♀ polled	—
Bardoka	Siberia, Yugoslavia	d.m.cw.	usually polled	—
Barki	Egypt	cw.m.d.	usually brown or black head; horned; polled; fat-tailed	—
Basque-Bearn	Basses-Pyrenees, France	d.m.cw.	♂ horned, ♀ polled	—
Beni-Ahsen	Northwest Morocco	cw.–mw.m.	white with colored face; horned or polled	—
Beni Guil	East Morocco, Algeria	m.cw.	white with colored head and legs; horned or polled	—
Berber	Morocco	m.cw.	usually white, also black or white with black head; horned	—

132

Bergamo	Italy	m.cw.	lop-eared; polled	1942
Bhadarwah	India	cw.pa.	face often colored; ♂ horned, ♀ polled; short-tailed	—
Bhakarwal	India	cw.m.	colored or white with color on face; long lop ears; ♂ horned; ♀ polled; some fat-tailed	—
Bibrik	Pakistan	cw.m.	usually black or brown muzzle; ♂ horned; ♀ polled; fat-tailed	—
Biella	Italy	m.cw.d.	lop-eared; usually white and polled; occasionally small horns in males	—
Bikaneri	India	cw.	face usually tan or brown; polled	—
Bizet	France	m.mw.	grey-brown with black on sides of face and on legs; horned; polled; long, thin tailed	—
Blackhead Persian	Africa	m.	hairy, woolless; polled; fat-rumped	1906
Black Merino	Portugal	fw.d.m.	brown to black; usually recessive ornamental dominant black	—
Black Welsh Mountain	—	—	polled	1922
Blanc du Massif Central	France	m.mw.d.	blue face, white hairs on black skin and finer fleece	—
Bluefaced Leicester	England	—	dark blue head and legs; polled	1963
Bluefaced Maine	France	m.lw.	polled	1938
Border Leicester	Scotland, England	lw.m.	white, often with spots on face and legs; occasionally colored; horned or polled	1897
Bosnian Mountain	Yugoslavia	d.m.cw.	—	—
Boulonnais	France	m.lw.	—	—
Bozakh	USSR	m.d.cw.	fat-tailed type: dirty-white, yellow-white, tan, grey, light red; usually polled	—
Braganca Galician	Portugal	m.cw.	white with colored spots on face, or black	—

(continued)

Table 7 (*Continued*)

Breed Name	Origin	Primary Use	Identifying Characteristics	Breed Society or Herd Book Established
Brazilian Woolless	Brazil	m.d.	red or white, occasionally pied; hairy, woolless; polled or horned	—
Buryat	USSR	m.cw.	white with black or red on head; ♂ horned, ♀ polled; short, fat tail	—
Calabrian	Italy	cw.d.m.	dark brown or white	—
Campanian Barbary	Italy	d.m.cw.-mw.	often dark spots on face and legs; horned or polled; short, fat tail	—
Campanica	Portugal	m.mw.d.	white, usually with brown spots on head and legs; ♂ horned, ♀ polled	—
Canadian Corriedale	Canada	mw.	—	—
Castilian	Spain	m.d.mw.	white, or black with white spot on head; polled	—
Caucasian	USSR	fw.m.	—	—
Central Pyrenean	France	m.mw.	—	—
Chanothar	India	cw.d.	light brown face; very long ears; polled	1927
Charmoise	France	m.sw.	polled	1936
Cher Berrichon	France	m.mw.	polled	—
Cherkasy	USSR	—	—	1893
Cheviot	England, Scotland	m.mw.	usually polled	—
Chios	Greece, Turkey	d.m.cw.-mw.	white with black or brown spots on face, belly, and legs; sometimes black face; horned or polled	—
Churro do Campo	Portugal	m.cw.	horned or polled	—
Chushka	USSR	fur, d.	black or white; horned	—
Clun Forest	England	sw.m.	dark brown face and legs; polled	1925
Columbia	United States	mw.m.	polled	—
Comiso	Sicily	d.cw.m.	reddish-brown face; polled	1942

Breed	Country	Code	Description	Year
Common Albanian	Albania	d.cw.m.	small, usually red or black spots on face and legs; horned or polled	—
Correidale	New Zealand	mw.m.	—	1910
Corsican	Corsica	d.m.cw.	white with black marks on head and feet; grey, or red; horned or polled	—
Cotentin	France	m.lw.	polled	1925
Cyprus Fat-tailed	—	d.cw.m.	fat-tailed; usually white with black or brown on face especially eyes and nose; occasionally black, brown, or white; usually horned or polled	—
Dagestan Mountain	USSR	m.mw.	—	—
Daglic	Turkey	cw.m.d.	often grey or black spots on head and legs; horned or polled	—
Dala	Norway	w.	polled	—
Dales-Bred	England	cw.	black-faced	1930
Dalmatian-Karst	Yugoslavia	m.cw.d.	usually white with white, black, or spotted head, also black or red-brown; horned or polled	—
Damani	Pakistan	d.cw.	brown markings on head; polled	—
Danube Merino	Romania	fw.d.m.	white, pied, or colored	1909
Dartmoor	England	m.lw.	black spots on nose; polled	—
Darvaz	USSR	cw.m.	black, white, or pied; horned or polled	—
Deccani	India	cw.	black, white, or pied; horned or polled	—
Degeres	USSR	m.	fat-rumped or short fat tail	1892
Derbyshire Gritstone	England	mw.m.	black-faced; polled	1923
Devon Closewool	England	sw.m.	polled	1898
Devon Longwooled	England	lw.m.	polled	1950
Dorper	Africa	m.	white with black head, often black feet; hairy, woolless, coarse-wooled; usually polled	
Dorset Down	England	sw.m.	brown face and legs; polled	1906
Dorset Horn	England	sw.m.	horned	1891

(continued)

135

Table 7 (*Continued*)

Breed Name	Origin	Primary Use	Identifying Characteristics	Breed Society or Herd Book Established
Doukkala	Morocco	cw.m.	usually white, head usually black or brown; horned or polled	—
Dubrovnik	Yugoslavia	mw.d.	white, black, or brown; polled	—
East Friesian	Germany	d.	polled; rat tail (woolless)	1892
Edilbaev	USSR	m.cw.	red or brown, occasionally black	—
Entre Minho e Douro	Portugal	mw.m.	usually white, black	—
Estonian Darkheaded	USSR	m.sw.	polled	—
Exmoor Horn	England	mw.m.	—	1907
Finnish Landrace	Finland	w.m.	short-tailed; usually white, occasionally black, brown, or grey; usually polled	1918
Frabosa	Italy	d.m.cw.	usually a few colored spots, sometimes brown; horned	—
French Alpine	France	m.mw.	polled	1952
French Blackheaded	France	m.sw.	black head and legs; polled	1959
Galway	Ireland	m.lw.	polled	1922
Garfagnana	Italy	d.m.w.	white, large; horned or polled	—
Georgian Finewool Fat-tailed	USSR	m.fw.	fat-tailed	—
Georgian Semifinewool	USSR	m.mw.	fat-tailed	—
German Blackheaded Mutton	Germany	m.sw.	polled	—
German Heath	Germany	—	coarse-wooled; grey with black face, lambs born black, horned, white horned, or polled	—
German Mutton Merino	Germany	m.fw.	polled	1947
German Whiteheaded Mutton	Germany	m.w.	polled	—

Breed	Origin		Code	Characteristics	Year
Gorki	USSR	—	m.sw.	brown face; polled	—
Greek Zackel		—	d.m.cw.	usually white with black or red spots on head and legs, occasionally colored; horned or polled	—
Grozny	USSR		fw.	—	—
Gujarati	India		cw.d.	brown face; polled	—
Gunib	USSR		cw.m.	fat-tailed; black; horned or polled	—
Gurez	India		cw.d.	usually white and polled; short-tailed	—
Hampshire Down	England		sw.m.	black-brown face and legs; polled	1889
Han-yang	China		mw.–cw.	white; polled; large and small fat-tailed	—
Harnai	Pakistan		cw.m.	colored spots on head; horned; fat-tailed	—
Hashtnagri	Peshawar, Pakistan, Afghanistan		cw.m.d.	black face; polled; fat-tailed	—
Hassan	India		cw.	white, black, and brown or grey; horned; polled	—
Hejazi	Aragia		m.	usually white; hairy, often woolless; often earless; polled or scurs; often tassels; short, fat tail	—
Herdwick	England		cw.m.	lamb black face and legs and blue roan fleece, adult white; ♂ horned, ♀ polled	1916
Hissar	USSR		m.cw.	brown; polled	
Hissar Dale	India		sw.	usually polled	
Hungarian Mutton Merino		—	fw.d.	—	
Hu-yang	China		cw.pelt	white; polled; short, fat tail	—
Icelandic		—	w.d.m.	short-tailed; usually white with light brown or pied; usually horned	—
Ile-de-France	France		m.mw.	polled	1922
Indre Berrichon	France		m.sw.	polled	1895
Iraq Kurdi	Iraq		m.cw.d.	black head and legs; polled; fat-tailed	—

(continued)

137

Table 7 (*Continued*)

Breed Name	Origin	Primary Use	Identifying Characteristics	Breed Society or Herd Book Established
Island Pramenka	Yugoslavia	cw.–mw.m.d.	usually white with white or speckled head and legs, occasionally black or brown; horned or polled	—
Istrian Milk	Yugoslavia	d.m.cw.	—	—
Jaidara	USSR	cw.m.	black, red, brown, or grey; horned or polled	—
Jalauni	India	cw.m.	face may have colored markings; usually polled; long lop ears	—
Kaghani	Pakistan	d.cw.	grey, brown, or black; lop ears; usually polled; short-tailed	—
Karabakh	USSR	m.d.cw.	fat-tailed type; usually reddish or greyish (dirty white or light brown), occasionally black or red; horned or polled	—
Karachaev	USSR	cw.m.d.	fat-tailed; 0–4 black horns	—
Karakachan	Macedonia	d.m.cw.	usually black or brown, occasionally white; horned or polled	—
Karakul	USSR	fur, d.	horned or polled; fat-tailed	1949
Karayaka	Turkey	cw.m.d.	usually white with black eyes or black head and legs; horned or polled; long, thin tail	—
Karnah	Pakistan	sw.	horned or polled	—
Karnobat	Bulgaria	d.w.	yellow-grey to black; horned	—
Kazakh Arkhar-Merino	USSR	m.fw.	horned, usually polled	—
Kazakh Fat-rumped	USSR China	cw.m.	reddish-brown; ♂ horned, ♀ polled	—
Kazakh Finewool	USSR	m.fw.	fat-rumped	—

Breed	Country		Description	Year
Kent or Romney Marsh	England	lw.m.	polled	1895
Kerry Hill	Wales	sw.m.	black eyes, nose, ears; polled; fat-tailed	1893
Khurasani	Iran	m.cw.d.	black spots on face and legs; polled	—
Kirgiz Fat-rumped	USSR	cw.m.	brown or black; horned or polled	—
Kirgiz Finewool	USSR	fw.m.	usually white	—
Kivircik	Turkey	m.d.cw.–mw.	white, with white or spotted face, also black or brown variant; horned or polled	—
Kosovo	Yugoslavia	m.cw.	black face and legs; usually polled	—
Krasnoyarsk	USSR	—	fine-wooled	—
Krivovir	Yugoslavia	m.cw.d.	white with brownish-yellow head and legs; occasionally brown, ♂ horned, ♀ polled	—
Kuche	China	pelt, m.cw.	black, pied, white, or brown; horned or polled; short-tailed to short, fat tail	—
Kuchugury	USSR	m.cw.	black with white patch on head; white polled	—
Kuibyshev	USSR	m.lw.	—	—
Kuka	Pakistan	cw.d.	face usually black; ♂ horned, ♀ polled, long lop ears	—
Lacaune	France	d.m.mw.	polled	1945
Lacho	Spain	d.cw.m.	long wool and black (brown or grey) face and feet; horned or polled	—
Lamon	Italy	cw.m.	lop-eared; dark spots on face and legs; polled	—
Langhe	Italy	d.m.cw.	horned or polled	1959
Latvian Darkheaded	USSR	m.sw.	polled	—
Lecce	Italy	d.cw.m.	white with black face and legs, occasionally black; ♂ horned, ♀ polled	1937
Leicester	England	lw.m.	polled	1893
Leine	Germany	w.m.	polled	1906

(continued)

139

Table 7 (*Continued*)

Breed Name	Origin	Primary Use	Identifying Characteristics	Breed Society or Herd Book Established
Lezgian	USSR	m.cw.	fat-tailed; white or greyish with colored or spotted head and feet, also black, red, grey, or pied	—
Libyan Barbary	Libya	m.cw.d.	white with brown or black (occasionally pied) face and legs, sometimes pied or colored; ♂ horned, ♀ polled; fat-tailed	—
Lika	Yugoslavia	d.m.cw.	white with colored or pied head and legs, sometimes brown or black	—
Limousin	France	m.mw.d.	polled	1906
Lincoln Longwool	England	lw.m.	polled	1892
Lipe	Yugoslavia	d.m.cw.	black head and legs; ♂ horned, ♀ polled	—
Liski	USSR	m.lw.	polled	—
Lithuanian Blackheaded	USSR	m.sw.	—	1934
Llanwenog	Wales	m.sw.	black head and legs; polled	—
Lohi	Pakistan	cw.m.d.	head black or brown; long lop ears usually with tassel; polled; short-tailed	—
Lonk	England	cw.m.	horned	—
Lot Causses	France	m.cw.	black spectacles and ears; polled	1955
Lourdes	France	m.mw.	white, occasionally brown or pied; ♂ horned, ♀ polled	—
Lowicz	—	lw.m.	polled	—
Macina	Niger, Mali	cw.	white, pied, or black; horned or polled	—
Malich	USSR	fur, d.	black, white, or grey; ♂ horned, ♀ polled; fat-tailed	—
Mancha	Spain	d.mw.m.	usually white, occasionally black; polled	—

Breed	Country	Use	Description	Date
Mandya	India	m.	hairy type; pale red-brown patches on anterior; usually polled; short-tailed	—
Manech	France	d.m.cw.	colored face and legs; horned	—
Masai	Tanzania, Kenya, Uganda	m.	fat-tailed; red-brown, occasionally pied; hairy, woolless; horned or polled	—
Massa	Italy	d.m.cw.	grey or brown with dark head; horned or polled	—
Maure	Africa	—	♂ horned, ♀ polled	—
Mazekh	USSR	m.d.cw.	fat-tailed; red-brown, black, red, or grey; horned, polled	—
Mikhnov	USSR	—	—	—
Miranda Galician	Portugal	m.cw.	white with spots on face and legs, or black	—
Mondego	Portugal	m.d.cw.	usually small brown spots on head and legs; horned	—
Mongolian	China	cw.m.d.	usually white with black or brown head, also pied or self-colored; ♂ horned, ♀ polled; fat-tailed	
Mytilene	Greece	d.m.cw.	usually white with spots on nose and legs; sometimes colored or pied; horned or polled; long, fat tail	—
Navajo	United States	cw.	horned or polled	—
Nejdi	Arabia	cw.	usually black with white head; polled or scurs; long, fat tail	—
Nellore	India	m.	hairy type; white or red-brown; ♂ horned, ♀ polled; tassels; short-tailed	—
New Zealand Romney Marsh	—	m.mw.	—	1904
North Caucasus Mutton Wool	USSR	m.mw.	—	—

(continued)

141

Table 7 (Continued)

Breed Name	Origin	Primary Use	Identifying Characteristics	Breed Society or Herd Book Established
North Country Cheviot	Scotland	—	—	1912
Northern Sudanese	Sudan	d.m.	long-legged; hairy, woolless; lop ears; long, thin tail	—
Old Norwegian	Norway	w.	short-tailed; usually white, sometimes colored; horned or polled	—
Oparino	USSR	m.w.	usually white with colored spots on head, neck, and legs, black or brown; usually polled	—
Ovce Polje	Yugoslavia	m.cw.d.	white, head and legs partly or wholly black or brown, occasionally black, brown, or grey; ♂ horned, ♀ polled	—
Oxford Down	England	sw.m.	dark brown face and legs; polled	1882
Pag Island	Yugoslavia	mw.d.	white, occasionally black; horned or polled	—
Pagliarola	Italy	cw.–mw.d.	yellowish-white, also reddish-black; ♂ horned, ♀ polled	—
Palas Merino	Romania	fw.m.	♂ horned, ♀ polled	—
Panama	United States	mw.m.	polled	—
Pirot	Yugoslavia	m.cw.	often spotted or colored head; ♂ horned, ♀ polled	—
Piva	Yugoslavia	m.cw.d.	usually white with spotted head and legs, occasionally black or grey	—
Pleven Blackhead	Bulgaria	d.cw.	grey with black head; ♂ horned, ♀ polled	—
Polish Heath	Poland	cw.pelt	—	—
Polish Merino	Poland	fw.m.	—	—
Polish Zackel	Poland	cw.d.pelt	white, also black or brown; horned, or polled	—

142

Breed	Country		Description	Year
Polwarth	Australia	mw.	—	—
Portuguese Merino	Portugal	fw.d.m.	—	—
Prealpes du Sud	France	m.sw.	polled	1948
Precoce	France	fw.m.	♂ horned or ♀ polled	—
Racka	Hungary	d.m.cw.	white or black	—
Radnor	Wales	m.sw.	tan or grey face and legs; horned or polled	1926
Rahmani	Egypt	cw.m.	brown, fading with age; often earless; ♂ horned, ♀ polled; fat-tailed	—
Rakhshani	Pakistan	cw.d.m.	black, white, or pied; ♂ horned, ♀ polled; fat-tailed	—
Rambouillet	France	fw.m.	♂ horned, ♀ polled	—
Red Karaman	Turkey	cw.m.d.	red, brown, or black; horned or polled	—
Rewshetilovka	USSR	—	♂ horned, ♀ polled	—
Rhon	Germany	w.m.	black head; polled	1921
Rila Monastery	Bulgaria	d.w.	♂ horned, ♀ polled	—
Romanov	USSR	pelt, cw.	grey with black head and legs and usually white face stripe and feet; horned or polled	—
Rough Fell	England	cw.	black faced; horned	1926
Russian Long-tailed	Russia	m.cw.	black or white; horned or polled	—
Russian Northern Short-tailed	USSR	pelt, cw.	black, grey, white, or pied; ♂ horned, ♂ polled	—
Ryeland	England	sw.m.	polled	1909
Rygja	Norway	sw.	face and legs sometimes colored; polled	—
Saloia	Portugal	d.mw.m.	usually white with pale brown head and legs, occasionally brown; ♂ horned, ♀ polled	—
Salsk Finewool	USSR	fw.	—	—
Saraja	USSR	cw.m.	grey; polled	—
Sardinian	Italy	d.m.cw.	occasionally black; horned or polled	1927
Sar Planina	Yugoslavia	m.cw.d.	♂ horned, ♀ polled	—

(continued)

143

Table 7 (*Continued*)

Breed Name	Origin	Primary Use	Identifying Characteristics	Breed Society or Herd Book Established
Savoy	Italy	d.m.cw.	black spots on face and legs; horned or polled	—
Scottish Blackface	Scotland, England	cw.m.	black or pied face; horned	1890
Serra da Estrela	Portugal	d.mw.m.	white, usually with brown spots on head and legs, black 7%	—
Shetland	Scotland	mw.m.	short-tailed; usually white, also moorit (brown), occasionally black, grey, or piebald; ♂ horned, ♀ polled	1926
Shkodra	Albania	cw.d.m.	yellowish face and legs; ♂ horned, ♀ polled	—
Shropshire	England	sw.m.	black-brown face and legs; polled	1882
Shumen	Bulgaria	—	colored	—
Sicilian	Italy	d.m.cw.	dirty-white, usually with black marks on face, occasionally black; ♂ horned, ♀ polled	—
Sicilian Barbary	Italy	cw.–mw.m.d.	usually dark spots on face and legs; ♂ horned, ♀ polled; long, fat tail	—
Sinkiang Finewool	China	fw.m.	fat-rumped	—
Sjenica	Yugoslavia	m.d.cw.	usually black around mouth, ears, and eyes; ♂ horned, ♀ polled	—
Skopelos	Greece	d.m.mw.	usually white with black or brown spots on face and legs; ♂ horned, ♀ polled	—
Sokolka	USSR	fur, d.	grey or black; horned	—
Solcava	Yugoslavia	m.cw.–mw.d.	lop-eared	—
Sologne	France	m.sw.	greyish with red-brown head and legs; polled	1948
Somali	Somalia, Ethiopia, Kenya	m.	white with black head; hairy, woolless; polled; fat-rumped	—

Breed	Country	Type	Description	Year
Sopravissana	Italy	fw.–mw.d.m.	♂ horned, ♀ polled	1942
South Africa Merino	—	fw.	—	—
South Devon	England	lw.m.	—	1904
Southdown	England	sw.m.	grey-brown face and legs; polled	1892
South Ural	USSR	fw.m.	♂ horned, ♀ polled	—
South Wales Mountain	—	—	tan face; bare belly; kempy (often red) fleece	—
Soviet Merino	USSR	fw.	—	—
Spanish Churro	Spain	d.m.cw.	black spots on face and feet; ♂ horned, ♀ polled	—
Spanish Merino	Spain	fw.	white, also black variant; polled	—
Stavropol	USSR	fw.	—	—
Steinschaf	Austria	m.cw.	small; white, black, grey, or brown; ♂ horned, ♀ polled	—
Stogos	Yugoslavia	m.cw.d.	white, usually with brown or yellow face and legs, occasionally brown or black; ♂ vertical screw horns, ♀ polled	—
Suffolk	England	sw.m.	black face and legs; polled	1886
Sumava	Czechoslovakia	cw.	white, usually with color on head and legs, also black or pied; ♂ horned, ♀ polled	—
Svishtov	Bulgaria	cw.	—	—
Svrljig	Yugoslavia	m.d.cw.	black on head and legs, occasionally all black; ♂ horned, ♀ polled	—
Swaledale	England	cw.m.	black face with grey-white muzzle; horned	1919
Swedish Landrace	—	w.m.	short-tailed; usually white, also pied or black	—
Swiss Black-Brown Mountain	Switzerland	sw.m.	polled	—
Swiss Brownheaded Mutton	Switzerland	m.sw.	polled	—

(continued)

145

Table 7 (*Continued*)

Breed Name	Origin	Primary Use	Identifying Characteristics	Breed Society or Herd Book Established
Swiss White Alpine	Switzerland	m.sw.	—	—
Swiss White Mountain	Switzerland	m.sw.	lop-eared	—
Tadla	Morocco	m.cw.	white with colored legs; horned	—
Tadmit	Algeria, Tunisia	m.mw.	♂ horned, ♀ polled	—
Tajik	USSR	w.m.	fat-rumped	—
Talavera	Spain	d.mw.m.	polled	—
Tanganyika Long-tailed	Tanzania	m.	fat-tailed; various colors; hairy, woolless; often earless and sometimes tassels; ♂ horned, ♀ polled; long thin tail	—
Tan-yang	China	cw.m.,pelt	white, usually with black or brown head and legs; horned, polled, or scurs; short, fat tail	—
Targhee	United States	mw.m.	polled	—
Teeswater	England	—	white or grey face	1949
Telengit	USSR	m.cw.	usually white with black or red head and neck, also black pied or red; short, fat tail	—
Texel	Netherlands	m.lw.	polled	—
Thal	Pakistan	cw.m.	pied, white, or black; polled; long lop ears	—
Thibar	Tunisia	mw.	black, occasionally with white spot on head or tail	1945
Thones-Marthod	France	m.cw.	black spectacles, nose tip, ears and feet; horned	—
Tibetan	China	cw.	usually black or brown head and legs; usually horned, short-tailed	—
Tirabi	Pakistan	cw.m.	fat-tailed	—
Transbaikal Finewool	USSR	fw.m.	—	—

Tsigai	Europe	mw.–cw.d.m.	dirty-white and black variants; white variant may have white face, red face, or black face; ♂ horned, ♀ polled	—
Tuareg	Sahara	m.	long-legged; white, pied, or fawn; ♂ horned, ♀ polled	—
Tuj	Turkey	m.cw.d.	sometimes dark marks around eyes and on feet; ♂ horned, ♀ polled; short, fat tail	—
Tung-yang	China	m.cw.–mw.	white; polled; tassels; large, fat tail	—
Tunisian Barbary	Tunisia	m.cw.	white with black or red-brown head, occasionally black or white; ♂ horned, ♀ polled; fat-tailed	1947
Turcana	Romania	d.cw.	black, grey, or white	—
Turkmen Fat-rumped	USSR	cw.	usually grey	—
Tushin	USSR	cw.m.d.	fat tail type	—
Tyrol Mountain	Austria	cw.m.	lop-eared; white face; long ears	—
Valachian	Czechoslovakia	d.m.cw.	white or black	—
Valais Blacknose	Switzerland	cw.m.	horned	—
Varese	Italy	m.cw.	lop-eared; polled	—
Velay Black	France	m.mw.	black with white spot on forehead and white tail tip; polled; long, thin tail	1931
Voloshian	USSR	m.cw.	white or black; horned or polled; long, fat tail	—
Vyatka	USSR	m.fw.	—	—
Waziri	Pakistan	cw.m.	usually light tan; long lop ears; fat-tailed	—
Welsh Mountain	—	w.m.	white or light tan face; ♂ horned, ♀ polled	1905
Wensleydale	England	lw.m.	blue face and legs; polled	1890
West African Dwarf	Africa	—	white, usually with red or black spots, occasionally black; hairy, woolless, horned	—

(continued)

147

Table 7 (*Continued*)

Breed Name	Origin	Primary Use	Identifying Characteristics	Breed Society or Herd Book Established
White Dorper	Africa	m.	all white variant of Dorper; hairy, woolless; usually polled	1960
White Face Dartmoor	England	m.lw.	occasionally speckled face; ♂ horned, ♀ polled	1950
White Karaman	Turkey	m.d.cw.	often black or brown marks on head; polled	—
White Klementina	Bulgaria	w.d.	—	—
White South Bulgarian	—	w.d.m.	—	—
Wicklow Mountain	Ireland	—	—	1943
Wiltshire Horn	England	m.	horned; woolless (i.e., very short, shedding fleece)	1923
Zante	Greece	m.cw.d.	white, occasionally with black spots on head; horned or polled	—
Zemmour	Morocco	m.cw.	white with pale brown face; horned, polled	—
Zeta Yellow	Yugoslavia	d.m.cw.	brownish-yellow head and legs	—

Abbreviations

     b. = bacon
   cw. = coarse-wooled (e.g., mattress, mixed, or carpet)
    d. = dairy (milk)
   dr. = draft
    fl. = fleece (mohair or down)
   fw. = fine-wooled (e.g., Merino type)
    h. = heavy (horse)
    l. = light (horse)
   ld. = lard
   lw. = long-wooled (medium fine)
   m. = meat
  mw. = medium wool (i.e., intermediate between lw. and fw.)
    p. = pork
  pa. = pack
  py. = pony
   ri. = riding
  sw. = short-wooled (medium fine)
   w. = wooled (but not cw. or fw.)

References

Felius, M. (1985). "Genus *Bos*: Cattle Breeds of the World." MSD AGVET, Div. of Merck and Company, Inc., Rahway, New York.

French, M. H., in association with Johansson, I., Joshi, N. R., and McLaughlin, E. A. (1966). European Breeds of Cattle. Vols. I and II. Food and Agricultural Organization of the United Nations, Rome.

Gray, A. P. (1972). "Mammalian Hybrids: A Check List with Bibliography," Technical Communication 10 (revised). Commonwealth Agricultural Bureaux, Farnham Royal Slough SL2, 3BN, England.

Mason, I. L. (1969). "A World Dictionary of Livestock Breeds: Types and Varieties," 2nd (revised) ed., Technical Communication 8. Commonwealth Agricultural Bureaux, Farnham Royal, Bucks, England.

Mason, I. L., and Maule, J. P. (1960). "Indigenous Livestock of Eastern and Southern Africa," Technical Communication No. 14. Commonwealth Agricultural Bureaux, Farnham Royal, Bucks, England.

Rouse, J. JE. (1970–1973). "World Cattle." Vols. I, II, and III. University of Oklahoma Press: Norman, Oklahoma.

A Compilation of Heritability Estimates for Farm Animals

R. H. Miller

U.S. Department of Agriculture
Agricultural Research Service
Beltsville, Maryland

Introduction

Genetic improvement of farm animals depends upon the existence of genetic variation. That is, to improve a productive characteristic, such as milk production of dairy cows, requires that cows must differ in the genes they possess that govern ability to secrete milk. There are two kinds of genetic variation: additive and nonadditive. Additive genetic variation refers to differences from animal to animal in those genes in which the changes in genetic ability due to replacement of one gene with another are additive; that is, there is a linear change in breeding value with the addition of successive allelic substitutions.

Nonadditive genetic variation is of two types: dominance and epistatic. Dominance variation arises when the change in genetic ability is nonlinear after the substitution of one allele for another at the same genetic locus. Epistatic variation arises when nonlinear changes in genetic ability occur at a locus due to a gene substitution at a different locus.

Heritability estimates reflect the degree to which additive gene effects exist in controlling variation in performance (estimates may also reflect some degree of epistatic variation, depending upon the method of estimation). When substantial additive variation exists, selection on phenotypic differences among animals will change the genetic ability of the animal population. When nonadditive variation is substantial, breeding methods of crossbreeding or crossing of different genetic lines will be effective in improving performance.

Heritability estimates fall between 0.0 and 1.0, apart from errors of estimation. A value of near 0.0 suggests that genetic differences have little or nothing to do with variation in performance. A heritability of near 1.0 suggests that environmental variation plays little part in variation in performance from animal to animal.

Certain patterns are expected in heritability estimates for various aspects of animal performance. Traits associated with reproduction are expected to have very low heritabilities (some exceptions occur, such as the Booroola gene in Australian Merino sheep, which has large effects on frequency of multiple births). For traits with little connection to ability of the animal to reproduce, such as rate of gain or body dimensions, higher heritabilities are expected. Actually, for traits with a long history of selection by man (such as milk production of the dairy cow), a gradual decrease in heritability is also theoretically expected, due to loss of genetic variation. However, this has seldom been observed, perhaps due to the existence of mechanisms that create new genetic variation, such as unequal crossing-over.

Tables 1–5 provide heritability estimates from the literature, resting heavily on standard textbook summaries.

Table 1
Beef Cattle

Trait	Estimate	Reference
Age of puberty	.20–.40	(Warwick and Legates, 1979)
Average daily gain (weaning)	.25–.30	(Johansson and Rendel, 1968)
Average daily gain (feedlot)	.45–.50	(Warwick and Legates, 1979)
Average daily gain (pasture)	.25–.30	(Warwick and Legates, 1979)
Birth weight	.40	(USDA, 1964)
Bone percentage	.50	(Johansson and Rendel, 1968)
Calving difficulty	.05–.15	(Warwick and Legates, 1979)
Calving rate	.15–.25	(Warwick and Legates, 1979)
Cancer eye	.30	(USDA, 1964)
Carcass grade	.30	(USDA, 1964)
Cow maternal ability	.40	(USDA, 1964)
Dressing percentage	.60	(Johansson and Rendel, 1968)
Fat thickness	.25–.40	(Warwick and Legates, 1979)
Feed conversion	.40	(USDA, 1964)
Feed intake	.25–.40	(Warwick and Legates, 1979)
Marbling score	.40–.60	(Warwick and Legates, 1979)
Multiple births	0.0–.10	(Warwick and Legates, 1979)
Palatability	.40–.70	(Warwick and Legates, 1979)
Rib-eye area	.25–.40	(Warwick and Legates, 1979)
Services/conception	0.0–.25	(Warwick and Legates, 1979)
Sperm count	.20–.30	(Warwick and Legates, 1979)
Tenderness	.60	(USDA, 1964)
Weaning weight	.30	(USDA, 1964)

Table 2
Dairy Cattle

Trait	Estimate	Reference
Lactation		
Fat percentage	.55–.70	(Warwick and Legates, 1979)
Fat yield	.20–.40	(Johansson and Rendel, 1968; Warwick and Legates, 1979)
Lactose percentage	.35	(Von Krosigk et al., 1960)
Lactose yield	.36	(Robertson et al., 1956)
Milk yield (305 day)	.20–.30	(Warwick and Legates, 1979)
Protein–lactose–mineral percentage	.50	(Warwick and Legates, 1979)
Protein–lactose–minerals	.20–.30	(Warwick and Legates, 1979)
Protein percentage	.50–.60	(Von Krosigk et al., 1960; Warwick and Legates, 1979)
Solids-not-fat percentage	.50–.60	(Robertson et al., 1956; Von Krosigk et al., 1960)
Solids yield (total)	.20–.30	(Warwick and Legates, 1979)
Milking/udder traits		
Average milking rate	.30–.40	(Warwick and Legates, 1979)
Height of fore udder	.08	(White, 1972)
Machine-on time	.22	(White, 1972)
Milking time	.02–.30	(Warwick and Legates, 1979)
Milk yield at 2 min	.35–.45	(Warwick and Legates, 1979)
Peak flow rate	.35–.45	(Warwick and Legates, 1979)
Percentage milk front quarters	.25–.35	(Pirchner, 1969)
Rear udder height	.09	(White, 1972)
Teat diameter	.38	(Johansson and Rendel, 1968)
Udder cleft	.14	(White, 1972)
Conformation		
Body capacity	.27	(Warwick and Legates, 1979)
Back	.23	(Warwick and Legates, 1979)
Dairy character	.19	(Warwick and Legates, 1979)
Feet	.11	(Warwick and Legates, 1979)
Fore udder	.21	(Warwick and Legates, 1979)
Final score	.31	
Front end	.12	(Johansson and Rendel, 1968)
General appearance	.29	(Warwick and Legates, 1979)
Head	.10	(Warwick and Legates, 1979)
Hind legs	.15	(Warwick and Legates, 1979)
Mammary system	.22	(Warwick and Legates, 1979)
Rear udder	.21	(Warwick and Legates, 1979)

(continued)

Table 2   (*Continued*)

Trait	Estimate	Reference
Rump	.25	(Warwick and Legates, 1979)
Stature	.51	(Warwick and Legates, 1979)
Teat placement	.31	(Warwick and Legates, 1979)
Udder support	.21	(Warwick and Legates, 1979)
Mastitis		
Cell count (log)	.14–.38	(Warwick and Legates, 1979)
Clinical frequency	0.0–.38	(Young *et al.*, 1960)
Nature of pathogens	.18	(Young *et al.*, 1960)
Fertility		
Abortion	.05	(Hendy and Bowman, 1970)
Calving interval	0.0–.10	(Warwick and Legates, 1979)
Cystic ovaries	.05	(Hendy and Bowman, 1970)
Days open	0.0–.09	(Warwick and Legates, 1979)
Gestation length	.50	(Pirchner, 1969)
Nonreturn rate	0.0–.03	(Warwick and Legates, 1979)
Nonreturn rate (sire mean)	.22	(Maijala, 1978)
Retained placenta	.16	(Pirchner, 1969)
Services/conception	0.0–.09	(Warwick and Legates, 1979)
Stillbirths	.01–.08	(Maijala, 1978)
Twinning	.11	(Hendy and Bowman, 1970)
Feed efficiency		
Feed efficiency (lactation)	.56	(Hooven *et al.*, 1972)
Feed efficiency (31–60 days)	.44	(Hooven *et al.*, 1972)
Feed efficiency (121–150 days)	.44	(Hooven *et al.*, 1972)
Feed efficiency (181–210 days)	.34	(Hooven *et al.*, 1972)
Feed intake (lactation)	.26	(Hooven *et al.*, 1972)
Feed intake (31–60 days)	.24	(Hooven *et al.*, 1972)
Feed intake (121–150 days)	.22	(Hooven *et al.*, 1972)
Feed intake (181–210 days)	.14	(Hooven *et al.*, 1972)
Growth/body size		
Average daily gain	.3–.5	(Warwick and Legates, 1979)
Birth weight	.40	(Warwick and Legates, 1979)
Chest depth	.38	(Colleau *et al.*, 1982)
Chest girth	.36	(Colleau *et al.*, 1982)
Chest width	.31	(Colleau *et al.*, 1982)
Feed conversion	.40	(Warwick and Legates, 1979)
Heart girth	.40–.60	(Warwick and Legates, 1979)
Mature size	.30–.50	(Warwick and Legates, 1979)
Rump length	.37	(Colleau *et al.*, 1982)
Width at hips	.38	(Colleau *et al.*, 1982)
Wither height	.50–.70	(Maijala, 1978)

Table 3
Poultry[a]

Trait	Estimate
Body weight	
2 wk (embryo)	.18
2 wk (male)	.01–.05
2 wk (female)	.31
4 wk (male)	.43
4 wk (female)	.31
6 wk (male)	.30
6 wk (female)	.38
8 wk (male)	.38
8 wk (female)	.39
10 wk (male)	.54
10 wk (female)	.51
12 wk (female)	.58
24 wk (pullets—light)	.57
24 wk (pullets—heavy)	.53
Mature (hens—light)	.52
Mature (hens—heavy)	.49
Weight gain	
Up to 10 wk (female)	.35
4–10 wk (male)	.44
Feed consumption	
Up to 10 wk (male)	.73
Up to 10 wk (female)	.87
Feed efficiency	
Up to 10 wk (male)	.14
Up to 10 wk (female)	.29
Egg production	
Short-term survivor production	.22
Intermediate survivor production	.19
Long-term survivor production	.22
Hen-housed production	
Short term	.33
Long term	.15
Rate of production	
Short term	.11
Long term	.15
Egg mass	.25

(continued)

Table 3 (*Continued*)

Trait	Estimate
Sexual maturity[b]	.39
Sexual maturity[c]	.39
Broodiness	.16
Pauses	.10
Persistency	.20
Oviposition interval	.59
Early egg weight[b]	.45
Early egg weight[c]	.57
Mature egg weight[b]	.46
Mature egg weight[c]	.58
Yolk weight	.43
Yolk size	.39
Albumen weight	.38
Blood spots	.13
Specific gravity[b]	.35
Specific gravity[c]	.40
Egg shape[b]	.35
Egg shape[c]	.32
Shell color	.35
Haugh units	.42
Miscellaneous traits	
Fertility	.02
Hatchability (percentage)	.09
Hatchability (probit)	.19
Hatching time	.49
Growing mortality	.02
Annual mortality	.08
Breast width	.17
Breast angle (male)	.39
Breast angle (female)	.40
Body depth	.19
Keel length	.47
Shank length	.13
Shank diameter	.61
Dressing percent	.41

Source: Kinney, 1969.
[a]Average of paternal half sib estimates. [b]Light breeds. [c]Heavy breeds.

Table 4
Sheep[a]

Trait	Estimate
Reproduction	
Fecundity	.10–.30
Semen quality	.05–.15
Number nipples	.22
Wool characters	
Grease fleece weight[b]	.30–.60
Clean fleece weight[b]	.30–.60
Percentage clean yield[c]	.10–.60
Staple length—weaning	.40–.50
Staple length—yearling	.40–.60
Fiber diameter[b]	.20–.50
Wrinkle score[c]	.20–.80
Growth	
Birth weight	.10–.30
Weaning weight	.10–.30
Yearling weight	.30–.40
Mature weight	.40–.60
Rate of gain	.30–.40
Gain/unit feed	.20–.40
Visual scores	
Weaning	.10–.30
Yearling	.10–.20
Face cover[b]	.40–.60
Carcass	
Weight	.30–.50
Conformation	.20–.30
Fatness	.15–.30
Loin eye area	.20–.30
Lean yield weight	.30–.50
Percentage lean	0.0–.1
Carcass (constant weight)	
Weight/day of age	.30–.50
Fatness	.40–.60
Quality grade	.30–.50
Loin eye area	.30–.50
Lean yield weight	.25–.40
Percentage lean	.25–.40

[a]All estimates from Warwick and Legates, 1979, except as indicated. [b]From Pirchner, 1969. [c]From Turner and Young, 1969.

Table 5
Swine

Trait	Estimate	Reference
Reproduction		
Birth weight	.09	(Hetzer and Miller, 1970)
Number farrowed	.05–.15	(Warwick and Legates, 1979)
Number weaned	.05–.15	(Warwick and Legates, 1979)
Weaning weight (pig)	.10–.20	(Warwick and Legates, 1979)
Age at puberty	.30–.40	(Warwick and Legates, 1979)
Gestation length	.41	(Hetzer and Miller, 1970)
Services per conception	0.0–.10	(Hetzer and Miller, 1970)
Carcass		
Dressing percentage	.25–.35	(Warwick and Legates, 1979)
Length	.40–.60	(Warwick and Legates, 1979)
Loin eye area	.40–.60	(Warwick and Legates, 1979)
Backfat thickness	.40–.60	(Warwick and Legates, 1979)
Belly thickness	.40–.60	(Warwick and Legates, 1979)
Percentage lean cuts	.40–.50	(Warwick and Legates, 1979)
Percentage of shoulder	.47	(Lasley, 1971)
Carcass score	.46	(Lasley, 1971)
Performance		
Average daily gain	.25–.40	(Pirchner, 1969)
Feed/unit gain	.30–.40	(Warwick and Legates, 1979)
Pig weight (140–180 days)	.20–.30	(Warwick and Legates, 1979)
Pig weight (98 days)	.15	(Hetzer and Miller, 1970)
Conformation		
Body length	.45–.60	(Pirchner, 1969)
Leg length	.65	(Lasley, 1971)
Number of ribs	.74	(Lasley, 1971)
Number of nipples	.20–.40	(Warwick and Legates, 1979)
Type	.38	(Lasley, 1971)

References

Colleau, J. J., L'Herminier, P., Tanguy, D., Felgines, C., and Le Mezec, P. (1982). Genetic parameters for type appraisal ratings in French Friesian cattle herd. *Proc. 2nd World Congress on Genetics Applied to Livestock Production* **VII**, 226–233.

Hendy, C. R. C., and Bowman, J. C. (1970). Twinning in cattle. *Anim. Breeding Abstracts* **38**, 22–37.

Hetzer, H. O., and Miller, R. H. (1970). Influence of selection for high and low fatness on reproductive performance of swine. *J. Animal Sci.* **30**, 481–495.

Hetzer, H. O., and Miller, R. H. (1972). Rate of growth as influenced by selection for high and low fatness in swine. *J. Anim. Sci.* **35**, 730–742.

Hooven, N. W., Miller, R. H., and Smith, J. W. (1972). Relationships among whole- and part-lactation gross feed consumption, and milk yield. *J. Dairy Sci.* **55**, 1113–1122.

Johansson, I., and Rendel, J. (1968). "Genetics and Animal Breeding." W. H. Freeman & Co., San Francisco.

Kinney, T. B. (1969). "A Summary of Reported Estimates of Heritabilities and of Genetic and Phenotypic Correlations for Traits of Chickens," Agriculture Handbook No. 363. Agricultural Research Service, USDA.

Lasley, J. (1971). "Genetics of Livestock Improvement," 2nd ed., Prentice-Hall, Inc., Englewood Cliffs, New Jersey.

Maijala, K. (1978). Breeding for improved reproduction in cattle. *Proc. Eur. Assoc. for Anim. Prod.* Stockholm, Sweden, June 5–7.

Philipsson, J. (1976). Studies on calving difficulty, stillbirth and associated factors in Swedish cattle breeds. III. Genetic parameters, *Acta Agric. Scand.* **26,** 211–220.

Pirchner, F. (1969). "Population Genetics and Animal Breeding." W. H. Freeman Co., San Francisco.

Robertson, A., Waite, R., and White, J. C. D. (1956). Variations in the chemical composition of milk with particular reference to the solids-not-fat. *J. Dairy Res.* **23,** 82.

Turner, H. N., and Young, S. S. Y. (1969). "Quantitative Genetics in Sheep Breeding." Cornell University Press, Ithaca, New York.

U.S. Department of Agriculture (1964). Beef cattle breeding. *Agr. Inf. Bull.* No. 286.

Von Krosigk, M., Young, J. O., and Richardson, G. A. (1960). Genetic influences on the composition of cow's milk. *J. Dairy Sci.* **43,** 877.

Warwick, E. J., and Legates, J. E. (1979). "Breeding and Improvement of Farm Animals." McGraw-Hill, New York.

White, J. M. (1972). Characteristics of the udder. Unpublished Proc. NC-2, S-49 Joint Meeting, Beltsville, Maryland.

Young, C. W., Legates, J. E., and Lecce, J. G. (1960). Genetic and phenotypic relationships between clinical mastitis, laboratory criteria, and udder height, *J. Dairy Sci.* **43,** 54–62.

Genetic Variation in Beef Cattle

Larry V. Cundiff
Keith E. Gregory
Robert M. Koch

Agricultural Research Service
U.S. Department of Agriculture
Clay Center, Nebraska
and
Department of Animal Science
University of Nebraska
Lincoln, Nebraska

Handbook of Animal Science

Introduction

Lush (1945) noted that stockmen have often been misled by the adage that "there is more variation within breeds than between breeds" into believing that genetic differences between breeds are "not real after all" or at least not very important. Therefore, it is important to assess the relative magnitude between and within breed genetic variation. Variation between breeds has accrued over long periods of time as a result of different selection goals, genetic drift (associated with inbreeding), or rare mutations. Heritability of breed differences is very high, provided the breed means are estimated with an adequate sample to average out errors of sampling individual animals within breeds. However, once between-breed genetic variation has been exploited by selection of breeds for use in an appropriate crossbreeding program, continued genetic improvement is dependent on intrapopulation genetic variation and selection. Heritability within breeds is lower than between breeds, but genetic variation within breeds is virtually restored generation after generation by the Mendelian process. Thus, selection within breeds should have the greatest long-term impact on genetic change for any specific objective. Although variation between breeds can only be exploited rarely when abrupt changes in selection goals are indicated and exotic germ plasm is available, opportunities to optimize performance levels by selection among breeds should not be overlooked.

Germ Plasm Evaluation Program

The relative amount of genetic variation between and within breeds for traits of economic importance to beef production has been evaluated in the Germ Plasm Evaluation (GPE) program at the Roman L. Hruska U.S. Meat Animal Research Center (MARC) (Cundiff *et al.*, 1986). Topcross performance of 20 different sire breeds has been evaluated in the first

Table 1
Sire Breeds in the MARC Evaluation Program[a]

	Cycle I (1970–72)	Cycle II (1973–74)	Cycle III (1975–76)	Cycle IV (1986–90)
F₁ crosses from Hereford or Angus dams (Phase 2)[b]	Hereford Angus Jersey South Devon Limousin Simmental Charolais	Hereford Angus Red Poll Brown Swiss Gelbvieh Maine Anjou Chianina	Hereford Angus Brahman Sahiwal Pinzgauer Tarentaise	Hereford[c] Angus[c] Longhorn Salers Galloway Nellore Shorthorn Piemontese Charolais Gelbvieh Pinzgauer
Three-way crosses out of F₁ dams (Phase 3)[d]	Hereford Angus Brahman Devon Holstein	Hereford Angus Brangus Santa Gertrudis		

[a]Germ Plasm Evaluation program at the Roman L. Hruska U.S. Meat Animal Research Center. [b]For example, Jersey sire × Hereford dam. [c]Hereford and Angus sires, originally sampled in 1969, 1970, and 1971, have been used throughout the program. In Cycle IV, a new sample of Hereford and Angus sires produced after 1982 are being used and compared to the original Hereford and Angus sires. Data on reproduction and maternal traits of females produced in Cycle IV are not available for presentation in this review. [d]For example, Brahman sire × (Jersey × Hereford) dam.

three of the four cycles of the GPE program (Table 1). In each cycle, Hereford–Angus reciprocal F₁ crosses (Hereford sires mated to Angus dams and vice versa) were produced using semen from the same sires throughout. Data presented were pooled over cycles by adding the average differences between Hereford–Angus reciprocal crosses (HA×) and other breed groups (two-way and three-way F₁ crosses) within each cycle to the average of Hereford–Angus reciprocal crosses (HA×) over the three cycles. Data on carcass and meat characteristics, obtained in cooperation with Kansas State University under the direction of Dr. Michael E. Dikeman, are presented for 15 F₁ crosses out of Hereford and Angus dams. Females produced by these matings were all retained to evaluate age and weight at puberty and reproduction and maternal performance through 7 or 8 yr of age. Data is presented for 19 F₁ crosses (two-way and three-way) grouped into seven biological types based on relative differences (X lowest, XXXXX highest) in growth rate and mature size, lean-to-fat ratio, age at puberty, and milk production (Table 2).

Table 2
Breed Crosses Grouped in Biological Types[a]

Breed Group	Growth Rate and Mature Size	Lean-to-Fat Ratio	Age at Puberty	Milk Production
Jersey (J)	X	X	X	XXXXX
Hereford-Angus (HA)	XX	XX	XXX	XX
Red Poll (R)	XX	XX	XX	XXX
Devon (D)	XX	XX	XXX	XX
South Devon (Sd)	XXX	XXX	XX	XXX
Tarentaise (T)	XXX	XXX	XX	XXX
Pinzgauer (P)	XXX	XXX	XX	XXX
Brangus (Bn)	XXX	XX	XXXX	XX
Santa Gert. (Sg)	XXX	XX	XXXX	XX
Sahiwal (Sw)	XX	XXX	XXXXX	XXX
Brahman (Bm)	XXXX	XXX	XXXXX	XXX
Brown Swiss (B)	XXXX	XXXX	XX	XXXX
Gelbvieh (G)	XXXX	XXXX	XX	XXXX
Holstein (Ho)	XXXX	XXXX	XX	XXXXX
Simmental (S)	XXXXX	XXXX	XXX	XXXX
Maine Anjou (M)	XXXXX	XXXX	XXX	XXX
Limousin (L)	XXX	XXXXX	XXXX	X
Charolais (C)	XXXXX	XXXXX	XXXX	X
Chianina (Ci)	XXXXX	XXXXX	XXXX	X

[a]On basis of four major criteria. Increasing number of X's indicate relatively higher values.

Variation between and within Breeds

Results for retail product growth to 458 days of age are summarized in Fig. 1. Retail product is closely trimmed boneless (trimmed to 8 mm of external fat and boneless except for dorsal and transverse spinous processes and rib bones in rib roasts) steaks, roasts, and lean trim. In Fig. 1, F_1 cross means for weight of retail product at 458 days of age are shown on the lower horizontal axis. The spacing on the vertical axis is arbitrary but the ranking of biological types (separate bars) from the bottom to top reflect increasing increments of mature size. Breed rankings within each biological type are noted within each bar. Steers sired by bulls of breeds with large mature size produced significantly more retail product than steers sired by bulls of breeds with small mature size.

In Fig. 1, differences are doubled in the upper horizontal scale to reflect variation among pure breeds relative to a standard deviation

VARIATION BETWEEN AND WITHIN BREEDS

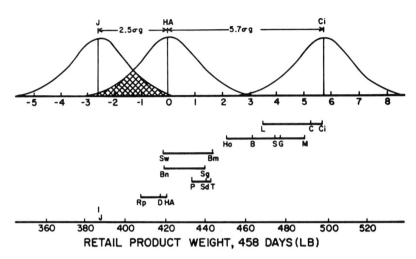

Figure 1 Breed group means (lower axis) and genetic variation between and within breeds (upper axis) for weight of retail product at 458 days. See Table 2 for abbreviations. (Adapted from Cundiff *et al.*, 1986).

change in breeding value [g = (²p) (h²)] within pure breeds for weight of retail product at 458 days of age. Frequency curves, shown for Jersey, the average of Hereford and Angus, and Chianina, reflect the distribution expected for breeding values of individual animals within pure breeds assuming a normal distribution (i.e., 68, 95, or 99.6% of the observations are expected to lie within the range bracketed by the mean ± 1, 2, or 3 standard deviations, respectively). The breeding value of the heaviest Jersey is not expected to equal that of the lightest Chianina, and the heaviest Hereford and Angus would only equal the lightest Chianina in genetic potential for retail product growth to 458 days. The range for mean differences between breeds is estimated to be about 5.7 σ_g between Chianina and Hereford or Angus steers and about 8.2 σ_g between Chianina and Jersey steers. Genetic variation, both between and within breeds, is considerable for this important measure of output. When both between- and within-breed genetic variations are considered, the range in breeding value from the smallest Jersey steers to the heaviest Chianina steers is estimated to be 180 kg, or 88% of the overall mean.

Retail product growth is not the only biological trait of economic importance to beef production that exhibits vast genetic variation, both between and within breeds. Cundiff *et al.* (1986) reported similar results

166   L. V. Cundiff, K. E. Gregory, and R. M. Koch

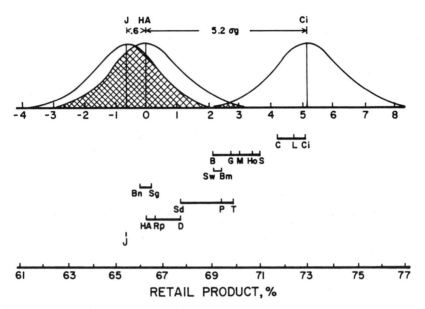

Figure 2 Breed group means (lower axis) and genetic variation between and within breeds (upper axis) for retail product as a percentage of carcass weight at 458 days of age. See Table 2 for abbreviations. (Adapted from Cundiff *et al.*, 1986.)

for other measures of growth (i.e., birth weight, weaning weight, post-weaning average daily gain, final slaughter weight, and carcass weight), gestation length, retail product percentage (Fig. 2), fat thickness, kidney fat percentage, marbling score (the primary determinant of USDA quality grade; Fig. 3), age of heifers at puberty (Fig. 4), milk production (Fig. 5), and mature cow weight (Fig. 6). The range for differences between breeds was at least comparable in magnitude to the range for breeding value of individuals within breeds for most traits evaluated.

Trade-offs

With so much genetic variation between breeds for retail product growth to a constant age, it is valid to ask why there hasn't been more done to exploit this variation. In the United States, Holsteins that excel in fluid milk yield have replaced the vast majority of cows of other breeds with lower genetic potential for fluid milk yield in dairy produc-

VARIATION BETWEEN AND WITHIN BREEDS

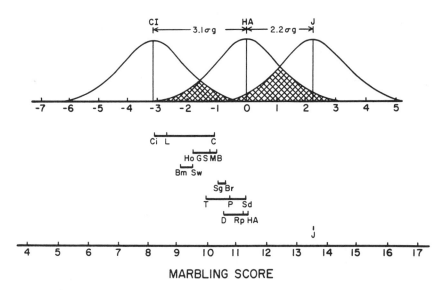

Figure 3 Breed group means (lower axis) and genetic variation between and within breeds (upper axis) for marbling score. See Table 2 for abbreviations. (Adapted from Cundiff *et al.*, 1986.)

tion. It is estimated that Holsteins produce 90% of the milk in the United States. In beef production in the United States, breeds that excel in output of retail product have not been substituted to nearly this extent for those with lower output potential. Why? In part, the answer lies in trade-offs that result from antagonistic genetic relationships between retail product growth and other traits important to efficiency of beef production.

Breeds that excel in retail product growth from birth to market age sire progeny with heavier birth weights, greater calving difficulty, reduced calf survival, and reduced rebreeding in dams; produce carcasses with lower marbling (Fig. 4) but very acceptable meat tenderness; tend to reach puberty at an older age (Fig. 5); and generally have heavier mature weight (Fig. 6). Heavier mature weight increases output per cow (i.e., slaughter weight of cull cows) but also increases nutrient requirements for maintenance. Thus, differences in output tend to be offset by input differences for maintenance and lactation so that differences in life cycle efficiency are generally small.

Because of trade-offs resulting from antagonistic genetic relationships among breeds, it would be very difficult for any one breed to

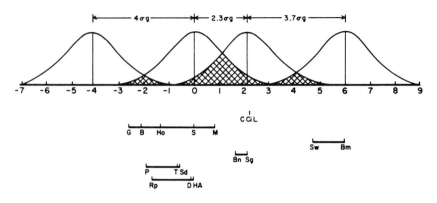

Figure 4   Breed group means (lower axis) and genetic variation between and within breeds (upper axis) for age of heifers at puberty (Adapted from Cundiff *et al.*, 1986.) A frequency curve is added to this figure to reflect variation between *Bos taurus* breeds only (i.e., Jersey to Charolais, Chianina, or Limousin) as well as that when both *Bos taurus* and *Bos indicus* breeds are considered (i.e., Jersey to Brahman). See Table 2 for abbreviations.

VARIATION BETWEEN AND WITHIN BREEDS

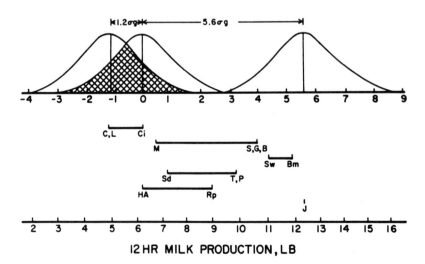

Figure 5   Breed group means (lower axis) and genetic variation between and within breeds (upper axis) for mean 12-hr milk production. See Table 2 for abbreviations. (Adapted from Cundiff *et al.*, 1986.)

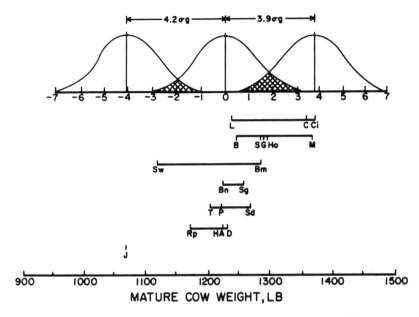

Figure 6   Breed group means (lower axis) and genetic variation between and within breeds (upper axis) for cow weight at 7 yr of age. See Table 2 for abbreviations. (Adapted from Cundiff *et al.*, 1986.)

excel in all characteristics of economic importance to beef production. Nor is it likely to be expected that simultaneous improvement in all characteristics occur from intrabreed selection since similar relationships exist within breeds. Use of crossbreeding systems that exploit complementarity by terminal crossing of sire breeds noted for lean tissue growth efficiency with crossbred cows of small to medium size and optimum milk production may provide the most effective means of managing trade-offs that result from genetic antagonisms.

References

Cundiff, L. V., Gregory, K. E., Koch, R. M., and Dickerson, G. E. (1986). Genetic diversity among cattle breeds and its use to increase beef production efficiency in a temperate environment. *Proc. 3rd World Congr. on Genet. Appl. to Livestock Prod. Lincoln, Nebraska* **IX,** 271.

Lush, J. L. (1945). "Animal Breeding Plans." Chap. 9. Iowa State University Press. Ames, Iowa.

Statistics

Populations

Paul A. Putnam

U.S. Department of Agriculture
Agricultural Research Service
Stoneville, Mississippi

Introduction

Population values have been selected and tabulated to illustrate comparison across major species of domestic livestock by country and continents. Changes in populations over time have also been included in tables 1, 2, 3, and 4. In tables 2, 3, and 4 when values were not available for the year cited, the adjacent year value was used if available. The entries in tables 2, 3, and 4 were limited to countries in which populations exceeded 5 million in two or more of the years cited. These countries represent the majority (approximately 80%) of the world popula-

Handbook of Animal Science

tion for the species reported. Table 5 includes world and continental populations for 1984, and Table 6 identifies countries having a million or more population of cattle, sheep, or hogs in 1984. Taken as a group, these tables provide perspectives on animal population trends across years among countries.

Table 1
U.S. Populations, 1900–1985[a]

Year	Humans[b]	Turkeys[c]	Chickens (All)[c]	Heifers and Milk Cows[c]	All Cattle[c]	Sheep and Lambs[c]	Hogs and Pigs[c]
1900	76	—	—	17	60	48	51
1905	84	—	—	18	66	44	53
1910	92	—	356	19	59	50	48
1915	101	—	379	20	64	41	57
1920	107	—	381	21	70	41	60
1925	116	—	435	23	63	39	56
1930	123	17	468	23	61	52	56
1935	127	20	390	26	69	52	39
1940	133	33	438	25	69	52	61
1945	133	42	516	28	86	47	59
1950	152	44	457	24	78	30	59
1955	166	65	391	23	97	32	50
1960	181	85	369	20	96	33	59
1965	194	106	394	17	109	25	51
1970	205	116	433	12	112	20	67
1975	216	124	380	11	132	15	49
1980	228	165	392	11	111	13	65
1985	239	171	374	11	110	10	54

[a]In millions. — indicates less than 1 million. [b]From U.S. Bureau of the Census, "Statistical Abstract of the United States: 1986." [c]From U.S. Department of Agriculture, Agricultural Statistics 1936/39– 1985.

Table 2
World Cattle, 1930–1985[a]

	1930[b]	1935[b]	1940[b]	1945	1950	1955	1960	1965	1970	1975	1980	1985
Africa												
Ethiopia	...	...	...	...	...	20.0	22.5	25.4	26.2	25.9	26.0	...
Kenya	—	5.2	5.2	—	—	6.9	7.5	7.7	8.6	9.7	10.0	...
Madagascar	7.2	6.4	5.2	5.9	5.7	6.1	6.4	8.5	9.9	8.7	10.2	...
Mali	...	...	...	...	...	...	—	—	5.4	—	5.9	...
Nigeria	...	...	...	...	...	...	...	11.1	11.2	11.0	12.3	...
South Africa	10.6	10.5	11.6	12.6	11.5	11.7	12.3	12.5	10.1	12.7	13.6	12.6
Sudan	...	...	...	...	6.0	7.0	9.1	13.8	14.7	18.4	...	...
Tanzania	...	...	...	...	...	...	7.9	10.0	13.2	11.3	12.6	...
Zimbabwe	...	...	...	...	...	...	...	...	6.1	5.3	...	...
Asia												
Bangladesh	...	...	...	...	...	...	...	...	26.8	25.5	33.0	...
Burma	...	...	...	...	...	—	5.3	6.4	7.0	7.4	8.5	...
China	23.0	23.0	25.0	19.8	—	44.8	44.5	62.8	63.2	56.7	52.5	...
India	190.4	204.7	180.0	136.4	133.8	158.7	158.7	175.9	176.5	179.5	182.5	249.7
Indonesia	...	...	7.6	...	...	5.0	—	6.5	6.1	6.2	6.4	...
Iran	—	—	...	...	...	5.0	5.4	5.3	5.2	7.2	7.2	...
Nepal	...	...	...	...	...	...	...	5.9	6.3	6.6	6.9	...
Pakistan	...	...	...	...	24.1	31.1	24.1	39.0	12.0	14.8	15.0	...
Thailand	8.8	10.0	11.3	...	5.0	5.7	5.3	5.2	—	...	...	...
Turkey	5.5	5.7	8.6	9.5	10.2	10.9	13.1	13.2	13.2	14.4	16.6	17.3
Europe												
France	14.9	15.6	15.5	14.0	15.4	17.3	18.7	20.2	21.7	24.0	23.9	23.1
Germany, East[c]	17.8	19.1	12.1	16.2	—	—	—	—	5.2	5.6	5.6	5.8
Germany, West	...	...	...	...	10.9	11.5	12.5	13.1	14.3	14.4	15.1	15.6

(continued)

Table 2 (*Continued*)

	1930[b]	1935[b]	1940[b]	1945	1950	1955	1960	1965	1970	1975	1980	1985
Ireland	—	—	—	—	—	—	—	5.4	6.0	7.2	6.9	5.8
Italy	7.3	7.2	8.2	6.6	8.3	8.7	9.4	9.2	9.6	8.2	8.7	9.0
Netherlands	—	—	—	—	—	—	—	—	—	—	5.2	5.3
Poland	9.0	9.5	9.9	—	6.8	7.9	8.7	9.9	10.8	13.3	12.6	11.1
Romania	—	—	—	...	...	—	—	—	—	5.8	6.3	7.0
United Kingdom	8.0	8.5	8.8	9.3	10.6	10.7	11.8	11.9	12.6	14.8	13.4	12.5
Yugoslavia	—	—	—	...	5.2	5.3	5.3	5.2	5.0	5.9	5.4	5.1
North America												
Canada	8.9	8.7	8.2	10.3	8.2	9.7	10.4	11.9	11.6	15.6	13.4	11.0
Cuba	—	—	—	—	...	...	5.8	6.6	7.1	5.5	5.9	...
Mexico	7.8	10.1	11.6	...	14.5	18.7	29.9	21.8	24.9	27.9	34.6	33.9
United States	59.2	68.4	68.2	85.6	78.0	96.6	96.2	109.0	112.3	132.0	111.2	108.8
Oceania												
Australia	11.9	12.9	13.3	14.1	14.6	15.8	16.5	18.8	22.2	32.8	26.2	22.5
New Zealand	—	—	—	—	5.0	5.9	6.0	6.8	8.8	9.3	8.1	7.9
South America												
Argentina	32.2	31.5	33.8	...	...	46.9	44.6	46.7	48.4	56.7	55.8	58.8
Brazil	47.5	44.2	40.8	44.6	52.1	61.4	72.8	84.2	75.4	92.5	119.0	93.7
Colombia	6.9	8.0	8.0	12.3	...	11.0	15.1	14.1	20.2	23.2	23.9	23.6
Paraguay	—	—	—	—	—	—	—	5.5	5.5	5.0	5.3	...
Uruguay	7.1	7.4	8.3	6.8	7.4	...	7.5	8.1	8.6	11.5	11.2	9.9
Venezuela	—	—	...	...	...	6.2	8.6	7.2	8.3	9.1	10.8	12.5
USSR	64.9	43.7	59.8	...	56.0	56.7	74.2	87.2	95.2	109.1	115.1	120.8

[a] In millions, for countries having more than 5 million. 1930–1940 and 1985 values are taken from USDA Agricultural Statistics HD 1751, UN 3a, 1936/39–1985. 1945–1980 values are taken from United Nations Statistical Yearbook HA 42, ST 29, Department of International Economical Social Affairs, Statistical Office, United Nations, New York, 1948–1983/84. ... indicates no information available. — indicates value less than 5 million. [b] Values are five-year averages. [c] Values cited for East Germany from 1930 to 1945 include both East and West populations.

Table 3
World Pigs, 1930–1985[a]

	1930[b]	1935[b]	1940[b]	1945	1950	1955	1960	1965	1970	1975	1980	1985
Asia												
China	80.0	84.0	65.0	48.5	59.5	87.9	180.0	202.0	223.0	263.6	325.1	306.1
India	—	—	—	...	—	...	—	5.0	—	7.2	8.2	...
Japan	—	—	—	...	—	—	—	—	6.3	7.7	10.0	10.8
Philippines	...	...	...	...	—	5.3	6.6	6.9	6.5	7.0	7.9	8.0
Vietnam	...	...	...	...	...	...	—	7.9	10.4	8.8	10.0	...
Europe												
Czechoslovakia	—	—	...	—	—	—	5.7	6.1	5.0	6.7	7.6	6.9
Denmark	—	—	...	—	—	—	6.1	8.6	8.4	7.7	10.0	9.0
France	5.9	6.6	7.0	—	6.8	7.6	8.4	9.0	10.5	12.0	11.4	10.1
Germany, East[c]	19.7	23.4	12.7	12.3	9.7	8.4	8.3	8.8	9.2	11.5	12.1	13.1
Germany, West	—	—	—	—	—	14.5	14.9	18.1	19.3	20.2	22.4	23.5
Hungary	—	—	...	—	6.5	5.8	5.4	7.0	5.9	8.3	8.4	9.8
Italy	—	—	—	...	—	—	—	5.4	9.2	8.8	8.8	9.0
Netherlands	5.7	6.5	—	...	—	—	—	—	5.7	7.3	10.1	11.3
Poland	—	—	—	...	7.7	10.9	12.6	13.8	13.4	21.3	21.3	17.2
Romania	—	—	—	...	—	—	—	6.0	6.0	8.6	10.9	14.8
Spain	5.0	5.1	—	—	5.7	6.0	6.0	—	6.9	7.9	11.3	12.0
United Kingdom	—	—	—	—	—	5.8	5.7	8.0	8.1	7.5	7.8	7.6
Yugoslavia	—	—	—	...	—	—	6.2	6.9	5.5	7.7	7.5	8.7
North America												
Canada	—	—	—	7.6	5.4	—	6.1	5.6	6.5	5.5	10.1	10.8
Mexico	—	—	5.0	...	...	8.2	10.2	9.5	10.3	13.2	16.9	12.3
United States	56.8	54.8	48.4	59.3	60.5	50.5	59.0	50.8	56.7	54.7	67.4	54.0
South America												
Brazil	22.1	23.4	23.2	24.3	24.2	35.6	46.8	58.7	30.8	35.2	34.2	33.0
USSR	21.0	15.6	32.3	...	27.1	30.9	53.4	52.8	56.1	72.3	73.9	77.8

[a]In millions, for countries having more than 5 million. 1930–1940 and 1985 values are taken from USDA Agricultural Statistics HD 1751, UN 3a, 1936/39–1985. 1945–1980 values are taken from United Nations Statistical Yearbook HA 42, ST 29, Department of International Economical Social Affairs, Statistical Office, United Nations, New York, 1948–1983/84. — indicates value less than 5 million. ... indicates no information available. [b]Values are five-year averages. [c]Values cited for East Germany from 1930–1945 include both East and West Germany populations.

Table 4
World Sheep, 1930–1985[a]

	1930[b]	1935[b]	1940[b]	1945	1950	1955	1960	1965	1970	1975	1980	1985
Africa												
Algeria	6.2	5.3	6.2	5.4	—	6.0	5.5	5.0	8.3	9.8	13.4	...
Ethiopia	...	...	...	...	...	18.0	19.9	11.8	12.7	23.1	23.3	...
Kenya	—	—	—	—	—	—	6.7	6.7	—	—	5.0	...
Mali	...	...	...	...	...	...	—	—	5.8	5.0	6.3	...
Morocco	8.4	8.0	10.0	10.9	10.4	15.4	12.0	12.0	15.2	14.8	14.2	...
Nigeria	...	...	...	...	8.4	...	...	7.5	8.1	10.1	11.7	...
Somalia	...	...	...	...	...	...	...	—	—	9.4	10.1	...
South Africa	43.1	44.1	39.9	37.9	31.4	37.0	38.2	42.1	34.3	31.0	31.6	29.9
Sudan	...	...	...	5.2	5.6	6.0	7.8	8.7	13.5	13.8	17.8	...
Asia												
Afghanistan	...	...	...	...	...	...	...	19.0	19.0	20.4	18.7	...
China	24.0	24.0	30.0	8.1	10.5	50.4	59.0	67.1	70.6	94.7	102.6	...
India	38.5	43.3	41.0	48.0	35.8	39.2	39.2	41.7	42.6	40.0	41.3	41.0
Iran	8.6	8.7	14.5	16.5	18.0	17.8	27.0	27.7	35.0	36.0	33.8	...
Iraq	—	—	7.1	6.5	...	10.0	9.2	11.0	13.1	11.5	9.5	...
Mongolia	...	...	...	...	...	...	...	13.1	12.6	14.5	14.4	...
Pakistan	...	...	8.0	...	6.4	6.1	8.1	12.9	15.3	17.5	26.2	...
Syria	...	...	...	—	—	—	—	5.4	6.1	5.8	9.3	...
Turkey	11.9	11.6	21.7	22.5	23.1	26.8	33.6	32.7	36.4	40.5	46.0	50.7

Europe												
Bulgaria	8.4	8.2	...	6.4	...	7.8	6.8	10.4	9.2	9.8	10.5	11.0
France	10.6	9.8	9.0	6.2	7.5	8.0	8.9	8.8	10.0	10.6	11.9	11.4
Greece	6.6	7.2	...	5.3	6.8	8.7	9.4	8.1	7.7	8.3	8.0	8.7
Italy	11.3	9.6	9.7	6.7	10.3	9.0	8.3	7.9	8.1	8.0	9.1	8.9
Romania	12.9	12.2	...	...	7.3	10.9	11.2	12.7	13.8	13.9	15.8	18.7
Spain	20.0	17.9	20.0	23.5	26.0	16.3	22.6	20.3	18.7	16.3	14.2	17.9
United Kingdom	24.7	26.0	26.3	15.9	20.4	22.9	27.9	29.9	26.1	28.4	31.4	23.7
Yugoslavia	7.8	8.5	...	—	10.0	12.0	11.4	9.4	9.0	8.2	7.4	7.7
North America												
Mexico	—	—	—	—	5.1	5.5	5.0	6.1	6.1	6.4	6.5	...
United States	45.6	53.2	51.5	46.5	30.7	31.6	33.2	25.1	20.3	14.5	12.7	10.4
Oceania												
Australia	103.3	111.4	112.6	105.4	112.9	130.8	155.2	170.6	180.1	151.7	136.0	144.0
New Zealand	27.5	28.8	31.3	34.0	31.5	39.1	47.1	53.7	60.3	55.3	68.8	70.3
South America												
Argentina	44.4	40.6	44.9	56.2	...	45.2	50.2	49.0	43.0	34.7	31.0	32.8
Bolivia	—	5.2	—	...	—	6.5	6.2	6.1	6.4	7.7	9.1	...
Brazil	10.7	11.8	11.4	13.3	13.5	17.5	19.0	21.9	24.4	17.8	18.4	...
Chile	6.3	6.1	5.9	6.0	6.5	5.9	6.3	6.7	6.4	5.6	6.1	...
Peru	11.2	12.0	14.9	17.3	18.5	16.8	15.1	14.5	16.8	15.3	14.5	...
Uruguay	20.6	18.0	17.9	19.6	22.6	24.5	21.7	21.9	19.9	15.1	20.0	22.8
USSR	122.8	53.0	93.0	...	82.2	99.0	136.1	125.2	130.7	145.3	143.6	142.4

[a]In millions, for countries having more than 5 million. 1930–1940 and 1985 values are taken from USDA Agricultural Statistics HD 1751, UN 3a, 1936/39–1985. 1945–1980 values are taken from United Nations Statistical Yearbook HA 42, ST 29, Department of International Economical Social Affairs, Statistical Office, United Nations, New York, 1948–1983/84. ··· indicates no information available. — indicates value less than 5 million. [b]Values are five-year averages in most cases.

Table 5
World Livestock Populations, 1984[a]

	Cattle	Sheep	Pigs	Horses	Asses	Mules
World	1273	1140	787	64	40	15
Africa	176	189	11	4	12	2
Asia	374	323	367	17	19	5
Europe	134	146	181	5	1	—
North America	187	20	92	19	3	4
Oceania	31	210	5	1	—	—
South America	252	107	53	13	4	3
USSR	120	145	79	6	—	—

[a]Estimated to nearest million. — indicates value less than 1 million. From "United Nations Statistical Yearbook," 1983–84.

Table 6
Cattle, Sheep, and Hog Populations, 1984[a]

	Cattle	Sheep	Hogs
Africa			
Algeria	1.5	14.7	—
Cameroon	3.7	2.2	1.0
Chad	3.4	2.2	—
Egypt	1.8	1.5	—
Ethiopia	26.0	23.5	—
Kenya	12.0	6.7	—
Madagascar	10.4	—	1.4
Mali	6.0	6.3	—
Mauritania	1.3	5.0	...
Morocco	3.3	12.0	—
Namibia	2.0	6.0	—
Niger	3.5	3.5	—
Nigeria	11.8	12.8	1.3
Senegal	2.2	2.1	—
Somalia	3.6	9.7	—
South Africa	12.9	31.3	1.4
Sudan	19.6	20.0	...
Uganda	5.2	1.3	—
Tanzania	14.5	4.1	—
Asia			
Afghanistan	3.8	20.0	...
Bangladesh	36.3	2.0	...
Burma	9.6	—	2.8
China	58.6	98.9	304.4

(continued)

Table 6   (*Continued*)

	Cattle	Sheep	Hogs
India	182.2	40.9	8.7
Indonesia	6.8	4.8	3.6
Iran	8.2	34.0	—
Iraq	1.5	8.3	—
Japan	4.7	—	10.4
Kampuchea	1.5	—	1.0
Korea, North	1.0	—	2.7
Korea, South	2.0	—	3.6
Mongolia	2.4	14.1	—
Nepal	7.0	2.5	—
Pakistan	16.4	24.3	—
Philippines	1.9	—	7.8
Syria	—	14.0	. . .
Thailand	4.6	—	4.2
Turkey	17.2	48.7	—
Vietnam	2.0	—	11.2
Europe			
Austria	2.6	—	3.9
Belgium/Luxembourg	3.2	—	5.3
Bulgaria	1.8	11.0	3.8
Czechoslovakia	5.2	1.0	7.1
Denmark	2.9	—	9.0
Finland	1.6	—	1.4
France	23.6	12.3	11.4
Germany, East	5.9	2.4	13.1
Germany, West	15.6	1.2	23.4
Greece	—	8.5	1.3
Hungary	1.9	3.0	9.8
Ireland	6.8	3.8	1.1
Italy	9.1	9.2	9.2
Netherlands	5.5	—	11.0
Poland	11.2	4.5	16.7
Portugal	1.0	5.0	3.5
Romania	6.5	18.5	14.3
Spain	5.1	16.6	12.4
Sweden	1.9	—	2.7
Switzerland	1.9	—	2.0
United Kingdom	13.2	34.8	8.3
Yugoslavia	5.3	7.5	9.3
North America			
Canada	12.3	—	10.8
Cuba	6.4	—	2.3
Mexico	37.5	6.4	18.4
United States	114.0	11.4	55.8
Oceania			
Australia	22.2	139.2	2.5
New Zealand	7.9	70.3	—

(*continued*)

Table 6 (*Continued*)

	Cattle	Sheep	Hogs
South America			
Argentina	53.5	30.0	3.8
Bolivia	4.3	9.2	1.7
Brazil	132.8	17.5	33.0
Chile	3.9	6.3	1.2
Colombia	23.9	2.7	2.4
Ecuador	3.3	2.3	4.3
Paraguay	5.1	—	1.4
Peru	2.8	14.5	1.8
Uruguay	9.5	23.3	—
Venezuela	12.2	—	2.6
USSR	119.6	145.3	78.7

[a]In millions, in countries having more than 1 million in at least two of these species. — indicates less than 1 million. · · · indicates no information available. From "United Nations Statistical Yearbook," 1983–84.

Physiological Averages/Ranges

Michael J. Darre
Stephen A. Sulik
Donald M. Kinsman

Department of Animal Science
University of Connecticut
Storrs, Connecticut

Handbook of Animal Science Copyright © 1991 by Academic Press, Inc.
All rights of reproduction in any form reserved.

Introduction

This chapter represents a thorough search of the literature to compile heretofore scattered information on the major livestock and avian species used in modern animal production. Although not exhaustive in content, it represents information most used by animal scientists, teachers, researchers, and others interested in facts about animals. The information in tables 1–14 is meant to stand alone and be used for quick reference. The references cited can be used by the reader seeking more in-depth information on any particular species.

Table 1
Heart Rate, Blood Pressure, and Other Blood Parameters of Selected Species

Common Name	Species Name	Heart Rate (Beats/Min)	Blood Pressure (sys/dias mmHg)	Blood Volume (% live bw)	Blood pH	Hematocrit	Erythrocyte Diameter (u)
Beef cattle	Bos taurus	60–70[a]	134/88[b]	7.70[c]	7.38[b]	33–47[d]	5.6[e]
Dairy cattle	Bos taurus	60–70[a]	134/88[b]	7.70[f]	7.38[b]	32.4[g]	5.6[e]
Horse	Equus caballus	32–44[b]	80/50[b]	9.70[c]	7.35[b]	28–42[d]	5.6[e]
Goat	Capra hircus	70–135[a]	120/84[b]	7.00[c]	—	27–34[d]	4.1[e]
Swine	Sus scrofa	60–80[a]	169/108[b]	6.75[c]	7.44[b]	39.1[d]	6.2[e]
Sheep	Ovis aries	70–80[a]	114/68[b]	8.00[c]	7.35[b]	31.7[a]	5.0[e]
Rabbit	Oryctolagus cuniculus	123–304[h,a]	110/80[i]	5.50[j]	7.54[l]	41.5[a]	7.5[e]
Chicken	Gallus domesticus	178–458[k]	131/95[b]	5.00[l]	7.50[l]	32–45[l]	11.2 × 6.8[l]
Turkey	Meleagris gallopavo	160–288[k]	270/167[m]; 235/141[n]	6.00–8.00[l]		38[l]	15.5 × 7.5[l]

[a]Altman and Dittmer, 1964; Swenson, 1970a. [b]Altman and Dittmer, 1964. [c]Frandson, 1974; Cortise, 1943; Hansard et al., 1953. [d]Altman and Dittmer, 1964; Swenson, 1970b. [e]Swenson, 1970b. [f]Frandson, 1974; Cortise, 1943. [g]Swenson, 1970a. [h]American Association for Laboratory Animal Science, 1979. [i]Arrington and Kelly, 1976. [j]Fox et al., 1984; Cortise, 1943. [k]Altman and Dittmer, 1964; Sturkie, 1976. [l]Sturkie, 1986a. [m]Adult male Bronze Turkey. [n]Adult male Beltsville Turkey.

Table 2
Metabolic Parameters of Selected Species

Common Name	Species Name	Respiratory Rate (br/min)	Lung Tidal Volume (ml)	Body Temp (°C)	Lower Critical Temperature (°C)	Upper Critical Temperature (°C)	Basal Metabolic Rate (KCal)[a]
Beef cattle	*Bos taurus*	12–20[b]	3450.0[b]	38.5[c]	−21[d]	27[e]	8220
Dairy cattle	*Bos taurus*	18–28[b]	3450.0[b]	38.5[c]	−21[d]	27[e]	8220
Horse	*Equus caballus*	8–16[b]	9000.0[f]	38.0[c]	—	—	6840
Goat	*Capra hircus*	10–20[g]	310.0[f]	39.0[c]	20[d]	28[e]	—
Swine	*Sus scrofa*	8–18[g]	—	39.0[c]	18[d]	32[e]	3580
Sheep	*Ovis aries*	10–20[g]	—	39.5[h]	−3; 9[e,i]	29; 32[e,i]	1875
Rabbit	*Oryctolagus cuniculus*	36–56[k]	15.8[f,k]	38.9[k]	—	—	190[k]
Chicken	*Gallus domesticus*	15–30[l]	40.0[l,m]	41.7[l]	18[n]	26[n]	110
Turkey	*Meleagris gallopavo*	25–35[l]	—	—	—	—	—

[a]Calculated as $70 \times BW^{.75}$. [b]Altman and Dittmer, 1964. [c]Romans *et al.*, 1985. [d]Curtis, 1981. [e]Yousef, 1985. [f]Swenson, 1970a; Altman and Dittmer, 1964. [g]Tenny, 1970. [h]American Association for Laboratory Animal Science, 1979; Romans *et al.*, 1985. [i]−3 for 100 mm fleece; 9 for 50 mm fleece. [j]Ewe shorn and full fleece. [k]American Association for Laboratory Animal Science, 1979. [l]Fedde, 1976. [m]Indicates respiratory capacity including air sacs. [n]Curtis, 1981.

Table 3
Visual Parameters of Selected Species

Common Name	Species Name	Visual Field (Degrees)	Binocular Vision (Degrees)	Color Perception[a]
Beef cattle	*Bos taurus*	330–360[b]	25–50[b]	Yes
Dairy cattle	*Bos taurus*	330–360[b]	25–50[b]	Yes
Horse	*Equus caballus*	330–350[b]	30–70[b]	Yes
Goat	*Capra hircus*	320–340[b]	20–60[b]	Yes
Swine	*Sus scrofa*	310[b]	30–50[b]	Yes
Sheep	*Ovis aries*	330–360[b]	20–50[b]	Yes
Rabbit	*Oryctolagus cuniculus*	330[b]	10–35[b]	Yes
Chicken	*Gallus domesticus*	300[c]	10–35[c]	Yes
Turkey	*Meleagris gallopavo*	300[c]	10–35[c]	Yes

[a]Based upon presence of cone cells. [b]Fedde, 1976. [c]Kare and Rogers, 1976.

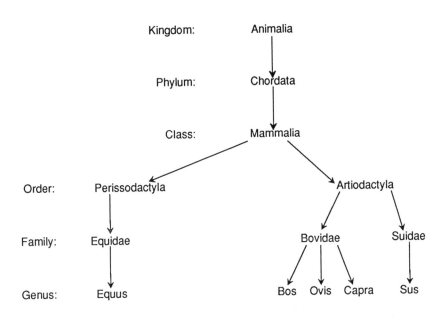

Figure 1   Zoological classification of farm animals.

Table 4
Organ Weights of Selected Species

Common Name	Species Name	Brain Weight (g)	Heart Weight (g)	Liver Weight (g)	Kidney Weight Right/Left (g)	Lung Weight (g)	Pancreas Weight (g)
Beef cattle[a]	Bos taurus	450.00	2230.0	5000	700/730	3500	350
Dairy cattle[a]	Bos taurus	450.00	2230.0	5000	700/730	3500	350
Horse[a]	Equus caballus	650.00	3400.0	5000	700/680	6000	350
Goat[a]	Capra hircus	—	—	—	—	—	—
Swine[a]	Sus scrofa	125.00	450.0	1000–2000	230/—	1000	—
Sheep[a]	Ovis aries	130.00	220.0	700	150/—	2500–3000	25–60
Rabbit	Oryctolagus cuniculus	13.21[b]	7.5[b]	105[b]	18.9[b]	9.1–10[b]	100–150
Chicken	Gallus domesticus	—	8.0[c]	—	41/—[c]	—	2–4[d]
Turkey	Meleagris gallopavo	—	—	—	—	—	—

[a]Frandson, 1974. [b]Fox et al., 1984; Arrington and Kelly, 1976. [c]Sturkie, 1976. [d]Hazelwood, 1986.

188

Table 5
Digestive Tract Lengths of Selected Species

Common Name	Species Name	Small Intestine Length (m)	Large Intestine Length (m)	Duodenum Length (m)	Cecum Length (cm)
Beef cattle	*Bos taurus*	27–49[a]	10.5[a]	1–1.2[b]	75[b]
Dairy cattle	*Bos taurus*	27–49[b]	10.5[b]	1–1.2[b]	75[b]
Horse	*Equus caballus*	19–30[b]	6.5[b]	1–1.5[b]	125[b]
Goat	*Capra hircus*	—	—	—	—
Swine	*Sus scrofa*	15–21[b]	3–4.5[b]	0.6–0.9[b]	20–30[b]
Sheep	*Ovis aries*	18–35[b]	4.5[b]	0.6–0.9[b]	18–35[b]
Rabbit	*Oryctolagus cuniculus*	3.56[c,d]	—	—	—
Chicken	*Gallus domesticus*	1.57[e]	.11[e]	.20[e]	.17[e]
Turkey	*Meleagris gallopavo*	—	—	—	—

[a]Frandson, 1974; Hill, 1970. [b]Frandson, 1974. [c]The combined length of the small and large intestine is approximately 11 times body length. [d]Hill, 1970. [e]Duke, 1986.

189

Table 6
Digestive Tract Capacities of Selected Species

Common Name	Species Name	Rumen–reticulum Capacity (l)	Omasum Capacity (l)	Abomasum Capacity (l)	Total Stomach Capacity (l)	Small Intestine Capacity (l)	Large Intestine Capacity (l)	Cecum Capacity (l)	Total Digestive Tract Capacity
Beef cattle	Bos taurus	125[a]	20[a]	15[a]	160[a]	65[a]	25[a]	10[a]	360[a]
Dairy cattle	Bos taurus	125[a]	20[a]	15[a]	160[a]	65[a]	25[a]	10[a]	360[a]
Horse	Equus caballus	—	—	8[a]	8[a]	27[a]	41[a]	14[a]	90[a]
Goat	Capra hircus	—	—	—	—	—	—	—	—
Swine	Sus scrofa	—	—	8[b]	8[b]	9[b]	9[b]	1[b]	27[b]
Sheep	Ovis aries	17[b]	1[b]	2[b]	20[b]	6[b]	3[b]	1[b]	30[b]
Rabbit	Oryctolagus cuniculus	—	—	—	—	—	—	—	—
Chicken	Gallus domesticus	—	—	—	3.2–6.4% of body weight/day[c]	—	—	—	—
Turkey	Meleagris gallopavo	—	—	—	—	—	—	—	—

[a]Maynard et al., 1979. [b]Kare and Rogers, 1976. [c]Duke, 1986.

Table 7
Dental Patterns and Vertebral Column Formulas
of Selected Species

Common Name	Species Name	Dental Pattern (Upper)[a]	Dental Pattern (Lower)[a]	Vertebral Column Formula[b]
Beef cattle[c]	*Bos taurus*	I0 C0 P6 M6	I8 C0 P6 M6	C7 T13 L6 S5 CD18–20
Dairy cattle[c]	*Bos taurus*	I0 C0 P6 M6	I8 C0 P6 M6	C7 T13 L6 S5 CD18–20
Horse[c]	*Equus caballus*	I6 C2 P6 M6	I6 C2 P6 M6	C7 T18 L6 S5 CD15–20
Goat[c]	*Capra hircus*	I0 C0 P6 M6	I8 C0 P6 M6	C7 T13 L7 S4 CD12
Swine[c]	*Sus scrofa*	I6 C2 P8 M6	I6 C2 P8 M6	C7 T14–15 L6–7 S4 CD20–23
Sheep[c]	*Ovis aries*	I0 C0 P6 M6	I8 C0 P6 M6	C7 T13 L6–7 S4 CD16–18
Rabbit[c]	*Oryctolagus cuniculus*	I4 C0 P6 M6–8	I2 C0 P4 M6	C7 T12 L7 S2–3 CDvar
Chicken	*Gallus domesticus*	—	—	C14 T7 LS14 CD6
Turkey	*Meleagris gallopavo*	—	—	—

[a] I = incisors; C = canines; P = premolars; M = molars. [b] C = cervical; T = thoracic; L = lumbar; S = sacral; LS = fused lumbar and sacral; CD = caudal or coccygeal. [c] Frandson, 1974.

Table 8
Mature Body Weights and Ages of Selected Species

Common Name	Species Name	Mature Body Weight (kg)	Mature Age	Longevity (Years)	Maximum Recorded Lifespan (yr)	Birth Weight (kg)
Beef cattle	Bos taurus	464–653[a]	5 yr[a]	20–25[b]	30[a]	25–45[a]
Dairy cattle	Bos taurus	464–653[a]	5 yr[a]	20–25[b]	30[a]	25–45[a]
Horse	Equus caballus	450[a,f]	3–5 yr[c]	20–25[c]	50[a]	45–55[a]
Goat	Capra hircus	26–102[a]	2 yr[a]	10–15[b]	18[a]	2.5–3.5[a]
Swine	Sus scrofa	70–128[a,d]	—	16–18[b]	27[a]	1.0–2.0[a]
Sheep	Ovis aries	34–80[a,d]	2 yr[a]	10–18[b]	20[a]	2.0–4.0[a]
Rabbit	Oryctolagus cuniculus	3.0–4.9[a]	1 yr[a]	5–8[b]	>13[a]	.04–.08[a]
Chicken	Gallus domesticus	1.5–3.3[a]	20–30 weeks[a]	2–5[a,e]	30[a]	.049–.065[a]
Turkey	Meleagris gallopavo	5.0–14.8[a]	28–35 weeks[a]	—	12.3[a]	0.52–.056[a]

[a]Altman and Dittmer, 1964. [b]Fox et al., 1984. [c]Blakely and Bade, 1982. [d]Cole and Garrett, 1980; Acker, 1983. [e]Depends upon use; commercial layers are usually within this range. [f]Ponies 227–409; light horses 409–636; draft horses 772–1000; from Ensminger, 1969.

Table 9
Reproductive Parameters of Selected Species

Common Name	Species Name	Age at Puberty—Males (Mo)	Age at Puberty—Females (Mo)	Length of Estrus (Hr)	Length of Estorus Cycle (Days)	Time of Ovulation	Gestation Period (Days)	Litter Size	Chromosome Number (2n)
Beef cattle	Bos taurus	10–12[b]	10–12[b]	18[c]	14–23[c]	25–30 hr FO[d]	270–290[c]	1–2[c]	60[d]
Dairy cattle	Bos taurus	10–12[b]	10–12[b]	18[c]	14–23[c]	25–30 hr FO[d]	270–290[b]	1–2[b]	60[d]
Horse	Equus caballus	12[e]	12–18[e]	120–168[c]	21–22[c]	24–48 hr BE[d]	337–344[c]	1–2[c]	64[d]
Goat	Capra hircus	8[f]	8[f]	48[c]	20–21[c]	—	147–155[c]	2–3[c]	60[d]
Swine	Sus scrofa	5–8[f]	6–8[f]	48–72[c]	18–24[c]	38–42 hr FO[d]	112–116[c]	6–12[c]	38[d]
Sheep	Ovis aries	7–8[f,g]	4–10[g]	36[c]	—	12–31 hr FO[d]	144–152[c]	1–3[c]	54[d]
Rabbit	Oryctolagus cuniculus	4–6[h]	4–6[h]	Constant[c]	—	10–12 hr AM[i]	30–31[c]	1–13[c]	44[j]
Chicken	Gallus domesticus	5–6[d,k]	5–6[d,k]	Constant[k]	—	22–26 hr BO[k]	21[k]	1/day[k]	78[k]
Turkey	Meleagris gallopavo	6–7[k]	6–7[k]	Constant[k]	—	22–26 hr BO[k]	21[k]	1/day[k]	80[k]

[a]FO = following onset of heat; BE = before estrus; AM = after mating; BO = before oviposition. [b]Foote, 1974. [c]Cole and Garrett, 1980. [d]Hafez, 1974. [e]Mishikawa and Hafez, 1974. [f]Altman and Dittmer, 1964. [g]Terrill, 1974. [h]Kraus et al., 1984. [i]Arrington and Kelly, 1976. [j]Kraus et al., 1984. [k]Nalbandov, 1979.

Table 10
Production Parameters of Selected Species[a]

Common Name	Species Name	Average Daily Gain (kg) (Prewean)	Average Daily Gain (kg) (Postwean)	% Product Recovery (Energy)	% Product Recovery (Protein)	kg/feed/kg Product
Beef cattle	Bos taurus	0.8[b]	1.15[b]	3–8[c]	4–15[c]	7.5[d]
Dairy cattle	Bos taurus	0.8[e]	1.15[e]	15–20[f]	15–36[f]	0.65 (milk)[d]
Horse	Equus caballus	—	1.4[g]	—	—	—
Goat	Capra hircus	—	—	—	—	88 (hair) / .48 (milk)
Swine	Sus scrofa	0.3[h]	1.4[h]	14–20[f]	—	3.82–3.94[d]
Sheep	Ovis aries	0.2[i]	.27[i]	—	14–20[f]	45 (wool)
Rabbit	Oryctolagus cuniculus	0.05[j]	—	—	—	3.0 (meat)[d]
Chicken	Gallus domesticus	—	0.05 (meat)[k]	6–11 (meat) / 10–18 (eggs)[f]	17–23 (meat) / 10–30 (eggs)[f]	2.5–4.8 (meat)[d] / 1.95 (meat)
Turkey	Meleagris gallopavo	—	—	—	—	2.6 (eggs)[l] / 3.0 (meat)

[a]All production parameters are estimates based upon normal feeding programs and healthy animals. [b]Maynard et al., 1979; National Research Council, 1976. [c]National Research Council, 1978a. [d]Acker, 1983. [e]Maynard et al., 1979; National Research Council, 1978a. [f]Maynard et al., 1979. [g]National Research Council, 1978b. [h]National Research Council, 1979. [i]Maynard et al., 1979; National Research Council, 1975. [j]Arrington and Kelly, 1976. [k]North, 1986. [l]Terrill, 1974.

Table 11
Energy Requirements of Selected Species

Common Name	Species Name	Energy Requirement (Growth) (kcal/day)[a]	Energy Requirement (Pregnancy) (kcal/day)[a]	Energy Requirement (Lactation) (kcal/day)[a]	Fecal Output (kg/day)	Urine Output (ml/day)
Beef cattle	*Bos taurus*	4200–31,000 ME[b]	12,600–24,000 ME[b]	15,900–27,500 ME[b]	13–35[c]	17–45/kg bw[d]
Dairy cattle	*Bos taurus*	1800–31,980 ME[e]	13,000–25,600 ME[e]	40,190–48,380 ME[e]	13–35[c]	17–45/kg bw[d]
Horse	*Equus caballus*	7430–19,200 DE[f]	8700–19,950 DE[g]	15,240–30,020 DE[g]	20[c]	—
Goat	*Capra hircus*	2100–3300 ME[h]	3100–4100 ME[h]	3400–4300 ME[h]	0.5–3.0[c]	10–40/kg bw[d]
Swine	*Sus scrofa*	2020–11,090 ME[i]	6430 ME[i]	15,840–17,420 ME[i]	0.5–3.0[c]	5–30/kg bw[d]
Sheep	*Ovis aries*	1980–2600 ME[j]	2160–4620 ME[j]	3580–6100 ME[j]	1.0–3.0[c]	10–40/kg bw[d]
Rabbit	*Oryctolagus cuniculus*	283 DE[k]	283–635 DE[k]	588	—	65/kg bw[l]
Chicken	*Gallus domesticus*	319 ME[m]	—	—	.11[c]	—
Turkey	*Meleagris gallopavo*	769 ME[m]	—	—	.45[c]	—

[a]ME = metabolizable energy; DE = digestible energy. [b]National Research Council, 1976. [c]Acker, 1983; Fox et al., 1984. [d]Fox et al., 1984. [e]National Research Council, 1978a. [f]Sturkie, 1986a. [g]National Research Council, 1978b. [h]Ensminger and Olentine, 1978. [i]National Research Council, 1979. [j]National Research Council, 1975. [k]National Research Council, 1977. [l]Kraus et al., 1984. [m]National Research Council, 1983.

Table 12
Water Requirements of Selected Species

Common Name	Species Name	Water Intake (Daily)[a]	Water Intake Requirement (Daily)[b]
Beef cattle	*Bos taurus*	57–170 l[c]	60 l[d]
Dairy cattle	*Bos taurus*	57–170 l[c]	90 l[d]
Horse	*Equus caballus*	2–4 l/kg dm[c]	50 l[d]
Goat	*Capra hircus*	—	—
Swine	*Sus scrofa*	1–3 l/kg dm[c]	>12 l[d]
Sheep	*Ovis aries*	1–3 l/kg dm[e]	>6 l[d]
Rabbit	*Oryctolagus cuniculus*	120 ml/kg[f]	—
Chicken	*Gallus domesticus*	2–3 g/g dm[e]	—
Turkey	*Meleagris gallopavo*	2–4 g/g dm[g]	—

[a]dm = dry matter; all dm values reflect approximate requirement per unit of dm intake. [b]Reflects approximate requirement during lactation. [c]Church, 1978. [d]Maynard et al., 1979. [e]National Research Council, 1983. [f]Kraus et al., 1984. [g]National Research Council, 1977.

Table 13
Body Composition as a Percentage of Live Weight of Selected Species

Common Name	Species Name	Muscle	Fat	Bone	Dressing	Hide	Blood	Viscera	Edible	Feet and Head	By-products
Beef cattle[a]	Bos taurus	36	14	10	60	7	5	20	3	5	40
Dairy cattle[a]	Bos taurus	26	14	10	50	7	5	30	3	5	50
Horse[a]	Equus caballus	—	—	—	—	—	—	—	—	—	—
Goat[a]	Capra hircus	—	—	—	—	—	—	—	—	—	—
Swine[a]	Sus scrofa	34	24	12	70	4	4	12	3	7	30
Sheep[a]	Ovis aries	28	15	7	50	10	5	26	3	6	50
Rabbit[b]	Oryctolagus cuniculus	40–47	8	7–9	55–62	—	5.5–5.7	—	—	—	38–45

[a]Kinsman, 1986. [b]Arrington and Kelly, 1976.

Table 14

Dressing Percentages of Processed Poultry as Percentage of Live Weight

Common Name	Species Name	Sex	Hot Dressed	Heads	Legs	Carcass and Neck	Giblets			
							Heart	Liver	Gizzard	Total
Chicken	*Gallus domesticus*	Male	94.5	2.6	4.7	74.2	0.4	1.9	1.6	3.9
		Female	92.6	2.5	3.9	73.8	0.4	1.9	2.3	4.7
Turkey	*Meleagris gallopavo*	Male	89.9	3.7	3.1	75.8	0.4	1.3	1.4	3.2
		Female	92.8	2.8	2.6	77.0	0.4	1.3	1.5	3.2

Source: Agricultural Research Service, 1986.

References

Acker, D. (1983). "Animal Science and Industry," 3rd ed. Prentice Hall, New Jersey.

Agricultural Research Service (1986). "Guidelines for Establishing and Operating Broiler Processing Plants," agricultural handbook 581. U.S. Department of Agriculture, Washington, D.C.

Altman, P. L., and Dittmer, D. S. (1964). "Biology Data Book." Federation of American Societies for Experimental Biology, Washington, D.C.

American Association for Laboratory Animal Science (1979). "Manual for Laboratory Animal Techniques." Joliet, Illinois.

Arrington, L. R., and Kelly, K. C. (1976). "Domestic Rabbit Biology and Production." The University Presses of Florida, Gainesville.

Blakely, J., and Bade, D. H. (1982). "The Science of Animal Husbandry," 3rd ed. Reston Book Co., Virginia.

Church, D. C. (1978). "Livestock Feeds and Feeding." O and B Books, Corvallis, Oregon.

Cole, H. H., and Garrett, W. N. (1980). "Animal Agriculture: The Biology, Husbandry and use of Domestic Animals," 2nd ed. W. H. Freeman and Co., San Francisco, California.

Cortise, F. C. (1943). The blood volume of normal animals. *J. Physiology* **102**, 290.

Curtis, S. E. (1981). "Environmental Management in Animal Agriculture," Animal Environment Services, Mahomet, Illinois.

Duke, G. E. (1986). Alimentary canal: Anatomy, regulation of feeding and motility. *In* "Avian Physiology," 4th ed. (P. D. Sturkie, ed.), Chap. 11. Springer Verlag, New York.

Ensminger, M. E. (1969). "Horses and Horsemanship." Interstate Printers and Publishers, Danville, Illinois.

Ensminger, M. E., and Olentine, C. G., Jr. (1978). Feeding goats. *In* "Feeds and Nutrition-complete," 1st ed. (M. E. Ensminger and C. G. Olentine, Jr. eds.), Chap. 22. Ensminger Publishing Co., Clovis, California.

Fedde, M. R. (1976). Respiration. *In* "Avian Physiology," 3rd ed. (P. D. Sturkie, ed.), p. 122. Springer Verlag, New York.

Foote, W. D. (1974). Cattle. *In* "Reproduction in Farm Animals," 3rd ed. (E. S. E. Hafez, ed.), Chap. 11. Lea and Febiger, Philadelphia, Pennsylvania.

Fox, J. G., Cohen, B. J., and Loew, F. M. (1984). "Laboratory Animal Medicine." Academic Press, New York.

Frandson, D. (1974). "Anatomy and Physiology of Farm Animals," 3rd ed. Lea and Febiger, Philadelphia, Pennsylvania.

Hafez, E. S. E. (1974). "Reproduction in Farm Animals," 3rd ed., p. 331. Lea and Febiger, Philadelphia, Pennsylvania.

Hansard, S. L., Butler, W. O., Comar, C. L., and Hobbs, C. S. (1953). Blood volume of farm animals. *J. Anim. Sci.* **12**, 402.

Hazelwood, R. L. (1986). Pancreas. *In* "Avian Physiology," 4th ed. (P. D. Sturkie, ed.), Chap. 23. Springer Verlag, New York.

Hill, K. J. (1970). Prehension, mastication, deglutition and the esophagus. *In* "Dukes' Physiology of Domestic Animals," 8th ed. (M. J. Swenson, ed.), Chap. 17. Cornell University Press, Ithaca, New York.

Kare, M. R., and Rogers, J. G., Jr. (1976). Sense organs. *In* "Avian Physiology," 3rd ed. (P. D. Sturkie, ed.). Springer Verlag, New York.

Kinsman, D. M. (1986). "Meat Animal Body Composition," handout 7/86. Department of Animal Science, University of Connecticut.

Kraus, A. L., Weisbroth, S. H., Flatt, R. E., and Brewer, N. (1984). Biology and diseases of rabbits. *In* "Laboratory Animal Medicine" (J. G. Fox, B. J. Cohen, and F. M. Loew, eds.), p. 207. Academic Press, New York.

Maynard, L. A., Loosli, J. K., Hintz, H. F., and Warner, R. G. (1979). "Animal Nutrition," 7th ed. McGraw-Hill, New York.

Mishikawa, Y., and Hafez, E. S. E. (1974). Horses. In "Reproduction in Farm Animals," 3rd ed. (E. S. E. Hafez, ed.), Chap. 14. Lea and Febiger, Philadelphia, Pennsylvania.

Nalbandov, A. V. (1979). "Reproductive Physiology of Mammals and Birds," 3rd ed. W. H. Freeman and Co., San Francisco, California.

National Research Council (1975). "Nutrient Requirements for Sheep." NAS-NRC, Washington, D.C.

National Research Council (1976). "Nutrient Requirements for Beef Cattle." NAS-NRC, Washington, D.C.

National Research Council (1977). "Nutrient Requirements of Rabbits." NAS-NRC, Washington, D.C.

National Research Council (1978a). "Nutrient Requirements for Dairy Cattle." NAS-NRC, Washington, D.C.

National Research Council (1978b). "Nutrient Requirements of Horses." NAS-NRC, Washington, D.C.

National Research Council (1979). "Nutrient Requirements of Swine." NAS-NRC, Washington, D.C.

National Research Council (1983). "Nutrient Requirements of Poultry." NAS-NRC, Washington, D.C.

North, M. O. (1986). "Commercial Chicken Production Manual," 3rd ed. AVI, Westport, Connecticut.

Pond, W. G., and Maner, J. H. (1974). "Swine Production in Temperate and Tropical Environments." W. H. Freeman and Co., San Francisco, California.

Romans, J. R., Jones, K. W., Costello, W. J., Carlson, C. W., and Ziegler, P. T. (1985). "The Meat We Eat," 12th ed. Interstate Printers and Publishers, Danville, Illinois.

Sturkie, P. D. (1976). Heart and circulation: Anatomy, hemodynamics, blood pressure, blood flow and body fluids. In "Avian Physiology," 3rd ed. (P. D. Sturkie, ed.), p. 76. Springer Verlag, New York.

Sturkie, P. D. (1986a). Body Fluids: Blood. In "Avian Physiology," 4th ed. (P. D. Sturkie, ed.), p. 102. Springer Verlag, New York.

Sturkie, P. D. (1986b). Kidneys: Extrarenal salt excretion and urine. In "Avian Physiology," 4th ed. (P. D. Sturkie, ed.), Chap. 16. Springer Verlag, New York.

Swenson, M. J. (1970a). "Dukes' Physiology of Domestic Animals," 8th ed. Cornell University Press, Ithaca, New York.

Swenson, M. J. (1970b). Physiologic properties, cellular and chemical constituents of blood. In "Dukes' Physiology of Domestic Animals," 8th ed. (M. J. Swenson, ed.), p. 21. Cornell University Press, Ithaca, New York.

Tenny, S. M. (1970). Respiration in mammals. In "Dukes' Physiology of Domestic Animals," 8th ed. (M. J. Swenson, ed.), Chap. 15. Cornell University Press, Ithaca, New York.

Terrill, C. E. (1974). Sheep. In "Reproduction in Farm Animals," 3rd ed. (E. S. E. Hafez, ed.), Chap. 12. Lea and Febriger, Philadelphia, Pennsylvania.

Yousef, M. K. (1985). "Stress Physiology of Livestock. I. Basic Principles," Chap. 7. CRC Press, Boca Raton, Florida.

Health

Internal and External Parasites

H. Herlich
M. D. Ruff

Agricultural Research Service
U.S. Department of Agriculture
Beltsville, Maryland

Introduction

Internal and external parasites affect all classes of food-producing animals (cattle, sheep, goats, swine, poultry), causing mortality and morbidity losses amounting to perhaps $1 billion annually in the United

Handbook of Animal Science Copyright © 1991 by Academic Press, Inc.
All rights of reproduction in any form reserved.

Table 1
Internal Parasites of Cattle

Name	Location	Transmission[a]
Protozoans		
Tritrichomonas foetus	Reproductive system	Venereal contact
Sarcocystis bovicanis (=cruzi)[b]	Striated muscle, heart	Ingestion of infective stages passed by dogs
Sarcocystis bovifelis[b]	Striated muscle, heart	Ingestion of infective stages passed by cats
Crystosporidum sp.	Small intestine	Direct
Toxoplasma gondii	All organs, muscles	Ingestion of infective stages passed by cats
Eimeria alabamensis	Small intestine	Direct
Eimeria auburnensis	Small intestine	Direct
Eimeria bovis	Small intestine, cecum, colon	Direct
Eimeria brasiliensis	Small intestine	Direct
Eimeria bukidnonensis	Small intestine	Direct
Eimeria canadensis	Small intestine	Direct
Eimeria cylindrica	Small intestine	Direct
Eimeria ellipsoidalis	Small intestine	Direct
Eimeria pellita	Small intestine	Direct
Eimeria subspherica	Small intestine	Direct
Eimeria wyomingensis	Small intestine	Direct
Eimeria zurnii	Small intestine, cecum, colon	Direct
Nematodes		
Dictyocaulus viviparus	Lungs	Direct
Gongylonema pulchrum	Esophagus	Ingestion of infected beetles, cockroaches
Gongylonema verrucosum	Rumen, abomasum	Probably ingestion of infected dung beetles
Haemonchus placei	Abomasum	Direct
Ostertagia ostertagi	Abomasum	Direct
Ostertagia lyrata	Abomasum	Direct
Trichostrongylus axei	Abomasum	Direct
Bunostomum phlebotomum	Small intestine	Direct; also by skin penetration
Cooperia oncophora	Small intestine	Direct
Cooperia pectinata	Small intestine	Direct
Cooperia punctata	Small intestine	Direct
Cooperia surnabada	Small intestine	Direct
Nematodirus helvetianus	Small intestine	Direct
Noeascaris (=Toxocara) *vitulorum*	Small intestine	Transmammary; perhaps transplacental and dire
Strongyloides papillosus	Small intestine	Direct; also by skin penetration
Trichostrongylus colubriformis	Small intestine	Direct
Trichostrongylus longispicularis	Small intestine	Direct

(continued)

Table 1 (*Continued*)

Name	Location	Transmission[a]
Oesophagostomum radiatum	Cecum, colon	Direct
Trichuris discolor	Cecum	Direct
Trichuris ovis	Cecum	Direct
Onchocerca lienalis	Ligaments of neck	Bite of black flies
Stephanofilaria stilesi	Skin (especially of abdomen)	Bite of black flies
Trematodes		
Fasciola hepatica	Liver	Ingestion of infective stage (metacercaria) in water or on grass; snail intermediate host required
Fasciola gigantica	Liver	Ingestion of infective stage (metacercaria) in water or on grass; snail intermediate host required (Hawaii only)
Fascioloides magna	Liver	Ingestion of infective stage (metacercaria) in water or on grass; snail intermediate host required
Dicrocoelium dendriticum	Liver	Ingestion of infected ants (second intermediate host); first intermediate host; snails
Paramphistomum microbothrioides	Rumen	Ingestion of infective stage (metacercaria) with water or grass
Paramphistomum cervi	Rumen	Ingestion of infective stage (metacercaria) with water or grass
Paramphistomum liorchis	Rumen	Ingestion of infective stage (metacercaria) with water or grass
Cestodes		
Moniezia benedeni	Small intestine	Ingestion of infected oribatid mites
Cysticercus bovis[a]	Heart, diaphragm, tongue, masseter; often liver, kidney, lungs	Ingestion of eggs passed by man
Cysticercus tenuicollis[a]	Mesentery, omentum, peritoneal surfaces	Ingestion of eggs passed by carnivores
Eichinococcus granulosus[a]	Lung, liver, brain, other organs	Ingestion of eggs passed by canids (dog, wolf)
Eichinococcus multilocularis	Lung, liver, brain, other organs	Ingestion of eggs passed by canids, primarily fox

[a]Direct = infection passed from animal to animal of same species by oral route. [b]Cattle serve as intermediate host, harboring only the immature stages of the parasite.

Table 2
Internal Parasites of Sheep and Goats

Name	Location	Transmission[a]
Protozoans		
Sarcocystis ovicanis	Striated muscle, heart	Ingestion of infective stages passed by canids
Sarcocystis capricanis	Striated muscle, heart	Ingestion of infective stages passed by canids
Sarcocystis ovifelis (=tenella)	Striated muscle, heart	Ingestion of infective stages passed by cats
Cryptosporidium sp.	Small intestine	Direct
Toxoplasma gondii	All organs, muscles	Ingestion of infective stages passed by cats
Eimeria ovina	Small intestine	Direct
Eimeria crandallis	Small intestine	Direct
Eimeria parva	Small intestine	Direct
Eimeria ovinoidalis	Small intestine	Direct
Eimeria ahsata	Small intestine	Direct
Eimeria intricata	Small intestine	Direct
Eimeria faurei	Small intestine	Direct
Eimeria granulosa	Small intestine	Direct
Eimeria pallida	Small intestine	Direct
Eimeria punctata	Small intestine	Direct
Thelazia californiensis	Eye	Muscid flies deposit infective larvae in secretions of eye
Dictyocaulus filaria	Lungs	Direct
Muellerius capillaris	Lungs	Ingestion of infected snail or slug
Protostrongylus rufescens	Lungs	Ingestion of infected snail or slug
Haemonchus contortus	Abomasum	Direct
Marshallagia marshalli	Abomasum	Direct
Ostertagia circumcincta	Abomasum	Direct
Ostertagia trifurcata	Abomasum	Direct
Trichostrongylus axei	Abomasum	Direct
Bunostomum trigono-cephalum	Small intestine	Direct; also by skin penetration
Cooperia curticei	Small intestine	Direct
Nematodirus abnormalis	Small intestine	Direct
Nematodirus filicollis	Small intestine	Direct
Nematodirus spathiger	Small intestine	Direct
Strongyloides papillosus	Small intestine	Direct; also by skin penetration
Trichostrongylus capricola	Small intestine	Direct
Trichostrongylus colubri-formis	Small intestine	Direct
Trichostrongylus longis-picularis	Small intestine	Direct

(*continued*)

Table 2 (*Continued*)

Name	Location	Transmission[a]
Trichostrongylus vitrinus	Small intestine	Direct
Oesophagostomum colum-bianum	Cecum, colon	Direct
Oesophagostomum venu-losum	Cecum, colon	Direct
Chabertia ovina	Colon	Direct
Trichuris ovis	Cecum	Direct
Trichuris skrjabini	Cecum	Direct
Elaeophora schneideri	Arteries, especially of head and face	Bite of horseflies
Cestodes		
Moniezie expansa	Small intestine	Ingestion of infected oribatid mites
Thysanosoma actinioides	Bile ducts, small intestine	Probably by ingestion of infected psocid mites
Cysticercus ovis[b]	Heart, diaphragm, muscle	Direct; eggs passed by infected carnivores (dog, fox)
Cysticercus tenuicollis[b]	Liver, peritoneal cavity	Direct; eggs passed by infected carnivores (dog, wolf, etc.)
Echinococcus granulosus[b]	Liver, lungs, and other organs	Ingestion of eggs passed by canids

[a]Direct = infection passed from animal to animal of same species by oral route. [b]Sheep and goats serve as intermediate host, harboring only the immature stages of the parasite.

States (ARS National Research Programs, 1976). Losses to endoparasites alone are estimated at $300 million annually for cattle (Tindell, 1983), $250 million for swine (Tindell, 1983), and $200 million for poultry (ARS National Research Programs, 1976). In addition to the direct effects on livestock, some internal parasites (e.g., *Cysticercus bovis, Cysticercus cellulosae, Trichinella spiralis, Toxoplasma gondii,* and *Sarcocystis* sp.) are important zoonotic agents (transmission to man) and are therefore public health problems (Steele, 1982). On the other hand, in addition to the direct effects exerted by ectoparasites on livestock, their greater importance undoubtedly is in their vector role—biologic or mechanical—in the transmission of viruses, rickettsias, and protozoans causing such devastating diseases as bluetongue, anaplasmosis, and babesiosis (Bram, 1978). Other losses to livestock producers are caused by condemnation of parts or entire carcasses because of parasites such as *C. bovis* and *T. spiralis* (75,000 entire swine carcasses in 1982) and of specific organs such as liver because of *Fasciola hepatica, Ascaris suum,* and *Stephanurus dentatus* (1.5 million beef livers in 1982 (U.S. Department of

Table 3
Internal Parasites of Swine

Name	Location	Transmission[a]
Protozoans		
Tritrichomonas suis	Nasal cavity, intestinal tract	Direct
Sarcocystis suicanis[b]	Small intestine	Ingestion of infective stage passed by canids (dog)
Sarcocystis porcifelis[b]	Small intestine	Ingestion of infective stage passed by cat
Cryptosporidium sp.	Small intestine	Direct
Toxoplasma gondii[b]	Small intestine	Ingestion of infective stage passed by cat
Eimeria debliecki	Small intestine	Direct
Eimeria cerdonis	Small intestine	Direct
Eimeria neodebliecki	Small intestine	Direct
Eimeria polita	Small intestine	Direct
Eimeria perminuta	Small intestine	Direct
Eimeria porci	Small intestine	Direct
Eimeria scabra	Small intestine	Direct
Eimeria suis	Small intestine	Direct
Eimeria spinosa	Small intestine	Direct
Isospora suis	Small intestine	Direct
Balantidium coli	Cecum, colon	Direct
Nematodes		
Gongylonema pulchrum	Esophagus	Ingestion of infected beetles and cockroaches
Ascarops strongylina	Stomach	Ingestion of infective dung beetles
Hyostrongylus rubidus	Stomach	Direct
Physocephalus sexalatus	Stomach	Ingestion of infected dung beetles
Trichostrongylus axei	Stomach	Direct
Ascaris suum	Small intestine	Direct
Strongyloides ransomi	Small intestine	Direct; also by skin penetration; transmammary and prenatal
Trichinella spiralis	Adults in small intestine, larvae in muscles	Ingestion of infected carnivore meat
Oesophagostomum brevicaudum	Colon	Direct
Oesophagostomum dentatum	Colon	Direct
Oesophagostomum georgianum	Colon	Direct
Oesophagostomum quadrispinulatum	Colon	Direct
Trichuris suis	Cecum	Direct
Metastrongylus apri	Lungs	Ingestion of infected earthworms

(continued)

Table 3   (*Continued*)

Name	Location	Transmission[a]
Metastrongylus salmi	Lungs	Ingestion of infected earthworms
Metastrongylus pudendotectus	Lungs	Ingestion of infected earthworms
Stephanurus dentatus	Kidneys	Direct; also by skin penetration; ingestion
Cestodes		
Cysticercus cellulosae[b]	Muscle	Ingestion of eggs passed by infected humans
Echinococcus granulosus[b]	Lungs, liver	Ingestion of eggs passed by infected dog
Trematodes		
Fasciola hepatica	Liver	Ingestion of metacercaria with water or grass
Paragonimus westermanni	Lungs	Ingestion of infected crayfish
Acanthocephalan		
Macracanthorhynchus hirudinaceus	Small intestine	Ingestion of infected beetles

[a]Direct = infection passed from animal to animal of same species by oral route. [b]Swine serve as intermediate host, harboring only the immature stages of the parasite.

Agriculture, 1983). The impact of parasitism on livestock is vividly exemplified in the fact that in the United States annual sales of endo- and ectoparasiticides are some $175 million for use in ruminants and swine (Harvey, 1977) and $110 million for use in poultry (ARS National Research Programs, 1976; Reid and McDougald, 1981).

There are thousands of species of internal and external parasites of livestock and poultry, Major groups of endoparasites include the protozoa, nematodes (roundworms), cestodes (tapeworms), and trematodes (flukes). Some parasites, such as coccidia, are extremely host-specific, while others, such as *T. gondii*, infect a variety of hosts. Major ectoparasite groups include lice, ticks, mites, and flies. Only those parasites of greatest economic importance and those of lesser importance but commonly found in the United States are listed in tables 1–6 (modification of Becklund, 1964). More detailed information can be found in Levine (1978) and Hofstad (1978).

The effects that parasites produce on their hosts may range from the innocuous—no detectable or measurable detriment—to death; usually the effects fall some place in between. The severity of harmful effects is modulated by a variety of factors, some related to the parasite and others to the host. Different species differ inherently in pathogenicity;

Table 4
External Parasites of Cattle, Sheep, Goats, and Swine

Cattle	Sheep and Goats	Swine
Mites		
Chorioptes bovis	Demodex caprae	Demodex phylloides
Demodex bovis	Demodex ovis	Sarcoptes scabiei
Psorergates bos		
Psoroptes bovis	Psoroptes cuniculi	
Raillietina auris	Sarcoptes scabiei	
Sarcoptes scabiei		
Ticks		
Amblyomma americanum	Amblyomma americanum	Amblyomma americanum
Amblyomma maculatum	Amblyomma maculatum	Amblyomma maculatum
Boophilus annulatus	Dermacentor occidentalis	Dermacentor variabilis
Dermacentor albipictus	Dermacentor variabilis	Otobius megnini
Dermacentor occidentalis	Otobius megnini	
Otobius megnini		
Lice		
Haematopinus eurysternus		Haematopinus suis
Linognathus vituli	Linognathus pedalis	
Solenopotes capillatus	Linoganthus stenopsis	
Bovicola bovis	Bovicola caprae	
	Bovicola ovis	
Flies		
Crysops sp.	Chrysops hominivorax	
Cochliomyia hominivorax	Melophagus ovinus	
Haematobia irritans	Phormia oetrus ovus	
Hypoderma bovis	Phormia regina	
Hypoderma lineatum		
Musca autumnalis		
Phormia regnia		
Stomoxys calcitrans		
Tabanus sp.		

for example, the abomasal nematode parasite of cattle, *Ostertagia ostertagi*, is more pathogenic than *Trichostrongylus axei*; similarly among the coccidial intestinal parasites of chickens, *Eimeria tenella* is much more pathogenic than *Eimeria acervulina* in terms of mortality, while the reverse is true for weight depression and pigment loss. Even within species, there are strains or geographic isolate differences, most notably again among the poultry coccidia. The level of exposure—the number of infective organisms entering the host—is critical; generally the greater the number of infective stages ingested or injected, the more severe the resulting infection. Young animals are usually more severely affected than old ones, probably because of the greater nutritional and physio-

Table 5
Endoparasites of Poultry

Name	Location	Intermediate Host	Definitive Host
Protozoa			
Eimeria truncata	Kidney	None	Duck, goose
Haemoproteus columbae	Blood	Pigeon fly	Pigeon
Haemoproteus meleagridis	Blood	Unknown	Turkey
Haemoproteus nettionis	Blood	Biting midge	Duck, goose
Leucocytozoon caulleryi	Blood, tissues	Biting midge	Pigeon
Leucocytozoon marchouxi	Blood, tissues	Unknown	Pigeon
Leucocytozoon simondi	Blood, tissues	Black fly	Duck, goose
Leucocytozoon smithi	Blood, tissues	Black fly	Turkey
Plasmodium relictum	Blood	Mosquito	Chicken, pigeon, duck
Plasmodium (other species)[a]	Blood	Mosquito	Chicken, turkey
Sarcocystis rileyi	Muscle	Unknown	Chicken, duck
Toxoplasma gondii	Tissue	Chicken, turkey, duck	Cats
Trypanosoma (various species)	Blood	Black fly, mosquito, pigeon fly	Chicken, pigeon
Trichomonas gallinae	Mouth, esophagus, crop, liver, proventriculus	None	Chicken, turkey, pigeon
Eimeria acervulina	Upper small intestine	None	Chicken, turkey, pigeon
Eimeria brunetti	Lower small intestine, large intestine, ceca	None	Chicken
Eimeria hagani	Upper small intestine	None	Chicken
Eimeria maxima	Small intestine	None	Chicken
Eimeria mivati	Small intestine, large intestine, ceca	None	Chicken
Eimeria mitis	Small intestine, large intestine, ceca	None	Chicken

(continued)

Table 5 (Continued)

Name	Location	Intermediate Host	Definitive Host
Eimeria necatrix	Small intestine, ceca (oocysts)	None	Chicken
Eimeria praecox	Upper small intestine	None	Chicken
Eimeria tenella	Ceca, large intestine	None	Chicken
Eimeria adenoeides	Lower small intestine, large intestine, ceca	None	Turkey
Eimeria dispersa	Small intestine	None	Turkey (chicken, pheasant, quail)
Eimeria gallopavonis	Lower small intestine, large intestine, ceca	None	Turkey
Eimeria innocua	Small intestine	None	Turkey
Eimeria meleagridis	Lower small intestine, large intestine, ceca	None	Turkey
Eimeria meleagrimitis	Small intestine, large intestine, ceca	None	Turkey
Eimeria subrotunda	Small intestine	None	Turkey
Eimeria anseris	Small intestine	None	Goose
Eimeria labbeana	Small intestine, ceca	None	Pigeon
Eimeria colchici	Ceca	None	Pheasant
Eimeria phasiani	Small intestine	None	Pheasant
Hexamita columbae	Small intestine	None	Pigeon
Hexamita meleagridis	Small intestine	None	Chicken, turkey, pheasant
Tyzzeria anseris	Small intestine	None	Goose
Tyzzeria parvula	Small intestine	None	Goose
Tyzzeria perniciosa	Small intestine	None	Duck
Wenyonella philiplevinei	Lower small intestine, large intestine	None	Duck
Chilomastix gallinarum	Ceca	None	Chicken, turkey
Cochlosoma anatis	Ceca	None	Turkey, duck

212

Cryptosporidium parvum	Ceca, respiratory tract	None	Chicken
Endolimax (various species)	Ceca	None	Chicken, turkey, pheasant, duck, goose
Entamoeba (various species)	Ceca	None	Chicken, turkey
Histomonas meleagridis	Ceca, liver	Eggs of cecal worm *Heterakis gallinarum*	Chicken, turkey, pheasant
Histomonas wenrichi	Ceca	Eggs of cecal worm *Heterakis gallinarum*	Chicken, turkey, pheasant
Trichomonas anatis	Ceca, large intestine	None	Duck
Trichomonas anseri	Ceca, large intestine	None	Goose
Trichomonas gallinarum	Ceca, liver	None	Chicken, turkey, pheasant, goose
Tritrichomonas eberthi	Ceca	None	Chicken, turkey, duck
Nematodes			
Cyathostoma bronchialis	Trachea	None or earthworms	Turkey, duck, goose
Syngamus trachea	Trachea	None	Chicken, turkey, pheasant, goose
Oxyspirura mansoni	Eye	Cockroaches	Chicken, turkey, pigeon, duck
Capillaria annulata	Esophagus, crop	Earthworms	Chicken, turkey, pheasant, goose
Capillaria contorta	Mouth, esophagus, crop	None or earthworms	Chicken, turkey, pheasant, duck
Gongylonema ingluvicola	Crop, esophagus, proventriculus	Beetles, cockroaches	Chicken, turkey, pheasant
Cyrnea colini	Proventriculus	Cockroaches	Turkey
Dispharynx nasuta	Proventriculus	Sowbugs	Chicken, turkey, pigeon, pheasant
Microtetrameres helix	Proventriculus	Grasshoppers	Pigeon
Tetrameres americana	Proventriculus	Grasshoppers, cockroaches	Chicken, turkey, pigeon, duck

(*continued*)

Table 5 (*Continued*)

Name	Location	Intermediate Host	Definitive Host
Tetrameres crami	Proventriculus	Amphipods	Duck
Tetrameres fissispina	Proventriculus	Amphipods, grasshoppers, cockroaches, earthworms	Chicken, turkey, pigeon, duck, goose
Amidostomum anseris	Gizzard	None	Pigeon, duck, goose
Amidostomum skrjabini	Gizzard	None	Duck, pigeon
Cheilospirura hamulosa	Gizzard	Grasshoppers, beetles	
Ascaridia columbae	Small intestine	None	Pigeon
Ascaridia dissimilis	Small intestine	None	Turkey
Ascaridia galli	Small intestine	None	Chicken, turkey, duck, goose
Capillaria anatis	Small intestine, cecum, cloaca	None	Chicken, turkey, pheasant, duck, goose
Capillaria bursata	Small intestine	Earthworms	Chicken, turkey, pheasant, goose
Capillaria caudinflata	Small intestine	Earthworms	Chicken, turkey, pigeon, pheasant, duck, goose
Capillaria obsignata	Small intestine, cecum	None	Chicken, turkey, pigeon, goose
Ornithostrongylus quadriradiatus	Small intestine	None	Pigeon
Heterakis dispar	Cecum	None	Duck, goose
Heterakis gallinarum	Cecum	None	Chicken, turkey, pheasant, duck, goose
Heterakis isolonche	Cecum	None	Pheasant, duck
Subulura brumpti	Cecum	Earwigs, grasshoppers, beetles, cockroaches	Chicken, turkey, pheasant, duck
Subulura stongylina	Cecum	Beetles, cockroaches, grasshoppers	Chicken
Strongyloides avium	Cecum	None	Chicken

214

Species	Location	Intermediate host	Definitive host
Trichostrongylus tenuis	Cecum	None	Chicken, turkey, pigeon, duck, goose
Capillaria columbae	Large intestine	None	Pigeon
Cestodes			
Amoebotaenia cuneata	Small intestine	Earthworms	Chicken, turkey, duck
Aporina delafondi	Small intestine	Unknown	Pigeon
Choanotaenia infundibulum	Small intestine	Housefly, beetles, grasshopper	Chicken, turkey
Davainea meleagridis	Small intestine	Unknown	Turkey
Davainea proglottina	Small intestine	Slugs, snails	Chicken
Diorchis nyrocae	Small intestine	Copepod	Duck
Drepanidotaenia watsoni	Small intestine	Unknown	Turkey
Fimbriaria fasciolaris	Small intestine	Copepods	Chicken, duck, goose
Hymenolepis anatina	Small intestine	Freshwater crustacea	Duck
Hymenolepis cantaniana	Small intestine	Beetles	Chicken, turkey, pheasant
Hymenolepis carioca	Small intestine	Beetles	Chicken, turkey
Hymenolepis collaris	Small intestine	Freshwater crustacea	Chicken, duck
Hymenolepis compressa	Small intestine	Unknown	Duck, goose
Hymenolepis coronula	Small intestine	Crustaceans, snails	Duck
Hymenolepis introversa	Small intestine	Unknown	Duck
Hymenolepis lanceolata	Small intestine	Crustaceans	Duck, goose
Hymenolepis megalops	Small intestine	Unknown	Duck
Hymenolepis parvula	Small intestine	Leeches	Duck
Hymenolepis tenuirostris	Small intestine	Copepods, amphipods	Goose
Imparmargo baileyi	Small intestine	Unknown	Turkey
Metroliasthes lucida	Small intestine	Grasshoppers	Chicken, turkey
Raillietina cesticillus	Small intestine	Beetles	Chicken, turkey
Raillietina echinobothrida	Small intestine	Ants	Chicken, turkey
Raillietina georgiensis	Small intestine	Ants	Turkey
Raillietina ransomi	Small intestine	Unknown	Turkey
Raillietina tetragona	Small intestine	Ants	Chicken, turkey
Raillietina williamsi	Small intestine	Unknown	Turkey

(continued)

215

Table 5 (*Continued*)

Name	Location	Intermediate Host	Definitive Host
Trematodes			
Collyriclum faba	Cysts in skin	Snail, dragonfly larvae	Chicken, turkey
Philophthalmus gralli	Eye	Snail	Chicken, turkey, duck, goose
Amphimerus elongatus	Liver	Snail, minnow	Chicken, turkey, duck
Prostogonimus macrorchis	Oviduct, bursa	Snail, dragonfly larvae	Chicken, turkey, duck
Renicola hayesannieae	Kidney	Snail, unknown	Duck, goose
Tanaisia brangai	Kidney	Snail	Chicken, turkey, pigeon
Trancheophilus cymbium	Respiratory system	Snail	Duck, goose
Typhlocoelum cucumerinum	Respiratory system	Snail	Duck, goose
Brachylaima fuscatum	Esophagus, small intestine, ceca	Snail	Chicken, pigeon, duck
Ribeiroia ondatrae	Proventriculus	Snail, fish, or tadpoles	Chicken, duck, goose
Apatemon gracilis	Small intestine	Snail	Duck, goose
Cotylurus cornutus	Small intestine	Snail	Turkey, pigeon, duck, goose
Cotylurus flabelliformis	Small intestine	Snail	Chicken, turkey, duck
Cryptocotyle concava	Small intestine	Snail, fish	Chicken, turkey, pigeon, duck
Echinoparyphium recurvatum	Small intestine	Snail, mollusk, or tadpoles	Chicken, turkey, duck

Species	Location	Intermediate host	Definitive host
Echinostoma revolutum	Small intestine, ceca, cloaca	Snail, mollusk, amphibians, fish	Chicken, turkey, pigeon, duck, goose
Hypoderaeum conoideum	Small intestine	Snail	Chicken, turkey, pigeon, duck, goose
Maritreminoides obstipus	Small intestine	Snail, isopod	Chicken, pigeon, duck
Paramonostomum parvum	Small intestine	Snail	Duck
Plagiorchis megalorchis	Small intestine	Snail, aquatic insects	Turkey, pheasant
Sphaeridiotrema globulus	Small intestine	Snail	Duck
Strigea falconis	Small intestine	Snail	Turkey
Brachylaima virginiana	Ceca	Snail	Turkey
Catatropis verrucosa	Ceca, bursa	Snail	Chicken, duck, goose
Notocotylus attenuatus	Ceca	Snail	Chicken, duck, goose
Notocotylus imbricatus	Ceca, cloaca	Snail	Chicken, duck
Postharmostomum gallinum	Ceca	Snail	Chicken, turkey, pigeon
Zygocotyle lunata	Ceca	Snail	Chicken, turkey, duck, goose
Acanthocephalans			
Oncicola canis	Tissue of esophagus	Turkey	Dog, coyote
Plagiorhynchus formosus	Small intestine	Unknown	Chicken
Polymorphus boschadis	Small intestine	Crustaceans, fish	Chicken, duck

[a]Severely pathogenic species (not found in the United States) include *P. gallinaceum* (chicken anbd pheasant), *P. juxtanucleare* (chicken, turkey), and *P. durae* (turkey).

Table 6
Arthropods on Poultry

Name	Location	Host
Mites		
Cytodites nudus	Air sacs	Chicken, turkey, pigeon, pheasant
Dermanyssus gallinae	Skin	Chicken, turkey, pigeon
Epidermoptes bilobatus	Skin	Chicken, turkey
Falculifer rostratus	Wing feathers	Chicken, turkey, pigeon
Freyana chaneyi	Shaft of wing feathers	Turkey
Knemidokoptes gallinae	Skin near base of feathers	Chicken, turkey, pigeon, pheasant
Knemidokoptes mutans	Legs, comb, neck	Chicken, turkey, pigeon, pheasant
Laminosioptes cysticola	Skin, muscles, body cavity	Chicken, turkey, pigeon, pheasant, goose
Megninia columbae	Neck and body feathers	Pigeon
Megninia cubitalis	Shaft of wing feathers	Chicken, turkey
Neonyssus columbae	Nasal cavity	Pigeon
Neonyssus melloi	Nasal cavity	Pigeon
Neoschongastia americana	Skin	Chicken, turkey
Ornithonyssus bursa	Feathers	Chicken, turkey, pigeon
Ornithonyssus sylviarum	Feathers	Chicken, turkey, pigeon
Pterolichus obtusus	Wing feathers	Chicken
Rivoltasia bifurcata	Feathers	Chicken
Speleognathus striatus	Nasal cavity	Pigeon
Syringophilus bipectinatus	Feather quill	Chicken, turkey, pheasant
Syringophilus columbae	Feather quill	Pigeon
Trombicula alfreddugesi	Skin	Chicken, turkey
Trombicula batatus	Skin	Chicken, turkey
Ticks		
Amblyomma americanum	External	Chicken
Argas persicus	External	All
Haemaphysalis chordeilis	External	Chicken, turkey
Haemaphysalis leporispalustris	External	Chicken, turkey
Ixodes brunneus	External	Turkey
Lice		
Amyrsidea megalosoma	Feathers	Pheasant
Anaticola adustus	Feathers	Goose
Anaticola anseris	Feathers	Goose
Anaticola crassicornis	Feathers	Duck
Anatoecus dentatus	Feathers	Duck, goose
Anatoecus icterodes	Feathers	Duck
Bonomiella columbae	Feathers	Pigeon
Campanulotes binentatus	Feathers	Pigeon
Ciconiphilus pectiniventrus	Feathers	Goose
Chelopistes meleagridis	Feathers	Goose, turkey
Coloceras damicorne	Feathers	Pigeon

(continued)

Table 6 (*Continued*)

Name	Location	Host
Colpocephalum tausi	Feathers	Turkey
Colpocephalum turbinatum	Feathers	Pigeon
Columbicola columbae	Feathers	Pigeon
Cuclotogaster heterographus	Head	Chicken, pheasant
Goniocotes chrysocephalus	Feathers	Pheasant
Goniocotes gallinae	Feathers	Chicken
Goniodes colchici	Feathers	Pheasant
Goniodes dissimilis	Feathers	Chicken
Goniodes gigas	Feathers	Chicken
Hohorstiella lata	Feathers	Pigeon
Holomenopon transvaalense	Feathers	Duck
Lagopoecus colchicus	Feathers	Pheasant
Lagopoecus sinensis	Feathers	Chicken
Lipeurus caponis	Wing feathers	Chicken, pheasant
Lipeurus lawrensis	Wing feathers	Chicken
Lipeurus maculosus	Feathers	Pheasant
Menacanthus cornutus	Body	Chicken
Menacanthus pallidulus	Body	Chicken
Menacanthus stramineus	Body	Chicken, turkey, pheasant, duck, goose
Menopon gallinae	Feathers	Chicken, pheasant
Ornithobius mathisi	Feathers	Goose
Oxylipeurus corpelentus	Feathers	Turkey
Oxylipeurus dentatus	Feathers	Chicken
Oxylipeurus mesopelios	Feathers	Pheasant
Oxylipeurus polytrapezius	Feathers	Turkey
Physconelloides zenaidurae	Feathers	Pigeon
Trinoton anserinum	Feathers	Goose
Trinoton querquedulae	Feathers	Duck
Bugs		
Cimex lectularius	Intermittent	Chicken, turkey, pigeon
Haematosiphon inodora	Intermittent	Chicken, turkey
Triatoma protracta	Intermittent	Chicken
Triatoma sanguisuga	Intermittent	Chicken, turkey, pigeon
Flies		
Aedes (various species)	Intermittent	All
Culex (various species)	Intermittent	All
Culicoides (various species)	Intermittent	All
Culiseta (various species)	Intermittent	All
Pseudolynchia canariensis	Intermittent	Pigeon
Psorophora (various species)	Intermittent	All
Simulium (various species)	Intermittent	All
Fleas		
Ceratophyllus gallinae	Intermittent	Chicken, pigeon
Ceratophyllus niger	Intermittent	Chicken, turkey
Echidnophaga gallinacea	Eyes, wattle, comb	Chicken, turkey, pigeon, pheasant

logical requirements of growing animals and because many young animals are not fully competent immunologically vis-à-vis specific parasites (Manton *et al.*, 1962). Furthermore, the young animal usually has not had sufficient time and exposure to develop an acquired immunity or is overwhelmed by infection before immunity can develop. Nutritional status is extremely important, as animals on subsistence or low protein rations and/or forage are generally more severely affected by parasitism (Reveron and Topps, 1970). Paradoxically, however, chickens on a low-level protein diet are less severely affected by the coccidia. The interaction of parasites and other diseases may be additive, mutually enhancing pathogenic effects on the host (e.g., coccidiosis and aflatoxin) (Wyatt *et al.*, 1975). Genetic variability of the host impacts on the host–parasite relationship. There are breed differences in resistance and susceptibility to parasitism (Bradley *et al.*, 1973; Drummond, 1975) as well as individual variability within breeds and species (Whitlock, 1958).

The direct effects of parasites on their hosts are many and varied, related in part to the hostal site parasitized, particularly in respect to internal parasites. Noticeable effects may include: anemia (pale mucous linings of eyes, lips, gums); hemorrhage (blood in feces); diarrhea (chronic or intermittent); constipation; inappetence; respiratory difficulty; wool shedding; feathering problems; decrease in milk and egg production; infertility problems (failure to conceive, tendency to abort, or decreased hatchability); weight loss; stunted growth; nervousness; and lethargy. More importantly, the most significant, although less dramatic, impact of parasitism, particularly in cattle, sheep, and poultry, is "subclinical" infection, where few if any of the listed effects are evident. In this form of infection animals continue to gain weight, grow, and produce but at less than optimal rates because of impaired digestion and malabsorption, causing great economic loss as a result of poor feed conversion ratios (Batte, 1974; Egerton, 1972; Drummond *et al.*, 1978).

Control of parasitism is largely effected through regular and judicious use of chemical parasiticides in conjunction with sound sanitary and management practices. Drug resistance in both internal and external parasites has been an increasing problem. Effective vaccines—with the exception of live cattle lungworm vaccine not available for use in the United States and a poultry vaccine for chickens but with limited use— have not been developed for any internal or external parasites. Much research is aimed at this method of control and some limited laboratory success has been achieved for some endoparasites of swine (Tromba, 1978), cattle (Herlich *et al.*, 1973), and sheep (Jarrett *et al.*, 1960). Biological control is an active area of research, the most notable success being the virtual eradication of the screwworm fly (*Cochliomyia hominivorax*) from the United States by controlled release of sterile flies (Knipling, 1979). Attempts to apply this technology to other arthropod pests of animals have not been practically successful.

References

ARS National Research Programs (1976). Swine (20430), Sheep (20440), Cattle (20420), Poultry (20450), and Insects Affecting Livestock (20480).

Batte, E. G. (1974). Advances in swine parasitology. In Proc. 22nd Pfizer Animal Research Conf., 119–135.

Becklund, W. W. (1964). Revised check list of internal and external parasites of domestic animals in the United States and possessions and in Canada. *Am. J. Vet. Res.* **25,** 1380.

Bradley, R. E., Sr., Radhakrishman, C. V., Patil-Kulkarni, V. G., and Loggins, P. E. (1973). Responses in Florida native and Romonillet lambs exposed to one and two oral doses of *Haemonchus contortus. Amer. J. Vet. Res.* **34,** 729.

Bram, R. A. (1978). "Surveillance and Collection of Arthropods of Veterinary Importance," USDA handbook 518.

Dineen, J. K., Gregg, P., and Lasceller, A. K. (1978). The response of lambs to vaccination at weaving with irradiated *Trichostrongylus colubriformis* larvae: Segregation into "responders" and "non-responders." *Int. J. Par.* **8,** 59.

Drummond, R. O. (1975). Tick-borne livestock diseases and their vectors. 4. Chemical control of ticks. *World Anim. Rev.* **16,** 28.

Drummond, R. O., Bram, R. A., and Konnerup, N. (1978). Animal pests and world food production. *In* "World Food, Past Losses, and the Environment" (AAAS Symposium 13), 63–93.

Egerton, J. R. (1972). The application of anthelmintics in the feedlot. *Adv. Pharmacol. Chemotherap.* **30,** 381.

Harvey, J. C. (1977). The market for external and internal parasiticides. *In* "Perspectives in the Control of Parasitic Diseases in Animals in Europe," p. 7. Royal College of Veterinary Surgeons, London.

Herlich, H., Douvres, F. W., and Romanowski, R. D. (1973). Vaccination against *Oesophagostomum radiatum* by injecting killed worm extracts and *in-vitro*-grown larvae into cattle. *J. Parasit.* **59,** 987.

Hofstad, M. S., Calnek, B. W., Hemboldt, C. F., Reid, W. M., and Yoder, H. W., Jr. (1978). "Diseases of Poultry," 7th ed. Iowa State University Press.

Jarrett, W. F. H., Jennings, F. W., and McIntyre, W. I. M. (1960). Resistance to *Trichostrongylus* produced by x-irradiated larvae. *Vet. Rec.* **72,** 884.

Knipling, E. F. (1979). "The basic principles of insect population suppression and management," USDA handbook 512.

Levine, N. D. (1978). "Textbook of Veterinary Parasitology." Burgess Publishing Co., Minneapolis, Minnesota.

Manton, J. V., Peacock, R., Poynter, D., Silverman, P. H., and Terry, R. J. (1962). The influence of age on naturally acquired resistance to *Haemonchus contortus* in lambs. *Res. Vet. Sci.* **3,** 308.

Reid, W. M., and McDougald, L. R. (1981). New drugs to control coccidiosis: To be or not to be. *Feedstuffs* **53,** 27.

Reveron, A. E., and Topps, J. H. (1970). Nutrition and gastrointestinal parasitism in ruminants. *Outlook Agriculture* **6,** 131.

Steele, J. H. (1982). "Handbook Series in Zoonoses, C. Parasite Zoonoses," Vols. 1–3. CRC Press, Boca Raton, FL.

Tindell, W., and Olentine, C. (1983). Parasites: Evaluating the problem. *Anim. Nutr. and Health* **39,** 24.

Tromba, F. G. (1978). Immunization of pigs against experimental *Ascaris suum* infection by feeding ultraviolet-attenuated eggs. *J. Parasit.* **64,** 651.

U.S. Department of Agriculture (1983). Statistical Summary. Federal Meat and Poultry Inspection for Fiscal Year 1982.

Whitlock, J. H. (1958). The inheritance of resistance to trichostrongylidosis in sheep. I. Demonstration of the validity of the phenomenon. *Cornell Vet.* **48,** 129.

Wyatt, R. D., Ruff, M. D., and Page, R. G. (1975). Interaction of aflatoxin with *Eimeria tenella* infection and monensin in young broiler chickens. *Avian Dis.* **19,** 730.

Major Infectious Diseases

O. H. V. Stalheim

Department of Veterinary Microbiology and
Preventive Medicine
Iowa State University
Ames, Iowa

Introduction

Diseases can be intrinsic (e.g., diabetes) or extrinsic. When the external agent (a virus, bacterium, or other microbe) is introduced in some way

Handbook of Animal Science   Copyright © 1991 by Academic Press, Inc.
All rights of reproduction in any form reserved.

Table 1
Diseases of Horses

Name	Hosts	Severity	Incidence	Distribution
African Horse Sickness	Horses, donkeys, elephants, dogs	Severe	Not in Western Hemisphere, Europe, Asia, Oceania	Africa; periodic epizootics in the Middle East
Babesiosis (piroplasmosis)	Horses	Inapparent to very severe	Enzootic	Southern United States, global
Colic	Equidae	Mild to severe	Sporadic	Global
Contagious equine metritis	Horses	Chronic	Rare; eliminated by treatment	United States, British Isles, France, Australia, Germany
Dermatophilosis	Horses, cattle, sheep, goats	Chronic	Sporadic	Global
Dourine	Horses	Chronic; venereal	Enzootic	South America, Middle East; not in United States
Eastern equine encephalomyelitis	Horses, donkeys, man, dogs, birds	Severe	Sporadic	Western Hemisphere; eastern United States only
Equine adenovirus	Horses	Mild to severe	Sporadic	Western Hemisphere, Europe, Australia

Disease	Host	Severity	Occurrence	Distribution
Equine arteritis	Horses	Inapparent to severe	Not known	Probably global
Equine coital exanthema	Horses only	Mild	Sporadic	North America, Europe
Equine ehrlichiosis (Potomac horse fever)	Horses	Mild to severe	Sporadic	Newly recognized
Equine infectious anemia	Horses, donkeys	Inapparent to severe	Endemic; diminished by testing	Global
Equine influenza	Equidae	Mild to severe	Epizootic	Global
Equine rhinopneumonitis	Horses	Mild to severe	Epizootic	Global
Equine rhinovirus	Horses	Mild to severe	Epizootic	North America, Europe
Glanders (farcy)	Horses	Severe	Now rare in developed countries	Not in United States
Periodic ophthalmia	Horses	Severe	Sporadic	Global
Ringworm	Most domestic animals, man	Mild	Sporadic	Global
Strangles	Horses	Mild to severe	Epizootic	Global
Surra	Equidae	Severe	Enzootic	Not in United States
Venezuelan equine encephalomyelitis	Horses, man	Mild to severe	Not in United States	Central and South America
Western equine encephalomyelitis	Horses, man, cattle, dogs, birds	Severe	Sporadic; diminished by vaccination	Western Hemisphere

Table 2
Diseases of Swine

Name	Hosts	Severity	Incidence	Distribution
African swine fever	Pigs, wart hogs	Severe	Not in United States	Africa; outbreaks in Iberia and the Caribbean
Atrophic rhinitis	Pigs	Mild to severe	—	Global
Coccidioidomycosis	Pigs, dogs	Mild	Sporadic	Southwestern United States
Encephalomyocarditis	Pigs, primates	Severe	Rarely in southern United States	Americas, Europe, Australia
Eperythrozoonosis	Pigs, cattle	Mild	Enzootic	Probably global
Erysipelas	Pigs, man	Severe	Common; diminished by vaccination	Global
Glasser's disease	Pigs	Mild to severe	Sporadic	Probably global
Hemagglutinating encephalomyelitis	Pigs	Mild to severe	Unknown	North America, Europe
Hog cholera	Pigs	Severe	Not in United States or Canada	Central and South America, Africa, Asia
Leptospirosis	Pigs, cattle, man	Mild to severe	Sporadic	Global
Mycoplasmal pneumonia	Pigs	Mild	Common	Global

Disease	Hosts	Severity	Occurrence	Distribution
Pasteurellosis	Pigs	Mild to severe	Sporadic	Probably global
Porcine cytomegalovirus	Pigs	Mild	Unknown	Global
Porcine parvovirus	Pigs	Severe *in utero*	Common	North America, Europe, Australia
Pseudorabies	Pigs	Inapparent to severe	Common; diminished by vaccination	Global except Canada
Streptococcal arthritis	Pigs	Mild to severe in young pigs	Epizootic	Global
Streptococcal lymphadenitis	Pigs	Mild	Sporadic	Global
Swine dysentery	Pigs	Mild to severe	Epizootic	Probably global
Swine influenza	Pigs, horses, man	Mild to severe	Epizootic	Global
Swine pox	Pigs only	Dermal only	Sporadic	Probably global
Swine vesicular	Pigs, sheep	Mild to severe	Not in United States	Central Europe
Teschen disease	Pigs	Severe	Unknown	Probably global
Transmissible gastroenteritis	Pigs	Severe in young pigs	Common	Probably global
Tuberculosis	Pigs, cattle, chickens	Mild	Sporadic	Global
Vesicular exanthema	Pigs	Mild	Not in United States	Probably eradicated

227

Table 3
Diseases of Cattle

Name	Hosts	Severity	Incidence	Distribution
Actinobacillosis	Cattle, swine, sheep	Chronic	Sporadic	Unknown
Actinomycosis	Cattle, sheep, dogs	Chronic	Sporadic	Unknown
Akabane virus	Ruminants, horses, monkeys, man	Mild to severe	Not in United States	Japan, Australia, Israel, and perhaps elsewhere
Anaplasmosis	Ruminants	Severe	Enzootic	North and South America
Anthrax	All warm-blooded animals	Severe	Enzootic, epizootic	Global
Astrovirus	Cattle, pigs man	Mild	Not in United States	Only Scotland so far
Bacillary hemoglobinuria	Cattle	Severe	Sporadic	Probably global
Blackleg	Cattle, sheep	Very severe	Sporadic; diminished by vaccination	Global
Bloat	Ruminants	Severe	Common	Global
Bovine genital campylobacteriosis	Cattle	Mild, venereal	Enzootic	Global
Bovine leukosis	Cattle	Mild to severe	Common in United States	Global except where eradicated
Bovine mammillitis	Cattle	Mild	Low	Africa
Bovine petechial fever	Cattle	Inapparent to fatal	Not in United States	Kenya
Bovine spongiform encephalopathy	Cattle	Severe	Newly recognized	United Kingdom
Bovine virus diarrhea	Cattle, sheep, goats	Mild to severe	Common	Global

Disease	Animals affected	Severity	Occurrence	Geographic distribution
Brucellosis	Cattle, swine, sheep	Mild, venereal	Diminished by testing and vaccination	Global
Contagious bovine pleuropneumonia	Cattle	Severe	Enzootic; not in United States	Africa, Asia
Cow pox	Cattle, cats, man	Mild to severe	Not in United States	Europe
Diarrhea (calf scours)	Young ruminants	Mild to severe	Common	Global
East Coast fever	Cattle	Severe	Enzootic; not in United States	Africa
Ephemeral fever	Cattle, buffalo	Mild to severe	Not in Western Hemisphere	Africa, Asia
Epidemic abortion	Cattle	Inapparent except for abortions	Low	Western United States
Foot and mouth	Cattle, all wild and domestic cloven-footed animals	Mild to severe	Not in North America	South America, Africa, Asia
Haemophilus septicemia	Cattle	Severe	Sporadic	Unknown
Infectious bovine rhinotracheitis	Cattle	Mild to severe	Common; diminished by vaccination	Global
Infectious keratitis	Cattle	Mild to severe	Common	Global
Ketosis	Cattle	Severe	Sporadic	Global
Leptospirosis	Cattle, sheep, pigs, dogs	Mild to severe	Common; diminished by vaccination	Global
Lumpy skin	Cattle, buffalo	Mild to severe	Not in United States	Africa only
Malignant catarrhal fever	Cattle, buffalo, sheep, deer	Severe	Sporadic	Global
Malignant edema	Cattle, horses, sheep	Very severe	Sporadic	Global

(continued)

229

Table 3 (*Continued*)

Name	Hosts	Severity	Incidence	Distribution
Mastitis	Cattle, sheep, goats	Mild to severe	Sporadic	Global
Mycoplasmal arthritis	Ruminants	Severe	Sporadic	Global
Nocardiosis	Cattle *et al.*	Chronic	Sporadic	Probably global
Paratuberculosis (Johne's Disease)	Ruminants	Severe	Enzootic	Probably global
Parturient paresis (milk fever)	Cows	Severe	Sporadic	Global
Pneumonic pasteurellosis	Cattle	Severe	Sporadic, epizootic	Probably global
Rinderpest	Cattle, sheep, goats, buffalo, pigs	Severe	Not in United States	Africa, Asia
Rotavirus	Cattle, pigs, man	Severe in neonates	Sporadic	Global
Russian spring–summer encephalitis	Cattle, goats, man	Severe	Not in United States	Asia, Europe
Trichomoniasis	Cattle	Mild, venereal	Sporadic	Global
Trypanasomiasis	Animals, man	Severe	Not in United States	Between latitudes 14° N and 29° S
Tuberculosis	Cattle, swine, chickens	Severe	Much diminished by testing	Global
Urinary calculi	Ruminants, dogs, cats	Severe	Sporadic	Global
Vesicular stomatitis	Cattle, horses, man	Inapparent to mild	Sporadic	North and South America

into an animal and multiplies, the animal has an infectious disease. If it spreads to other animals, it is considered a contagious disease. Consumers spend about $5 billion annually for veterinary services (Hayes, 1984), and a significant number of people (< 100) become ill each year with a zoonotic disease (i.e., one that is transmissible from animals to man) (U.S. Public Health Service, 1984). Depending upon the nature of the microbial agent, the route by which it enters an animal, the animal's innate or artificially induced immunity, and many other factors, the outcome of the infection can vary from inapparent to rapidly fatal. Some agents invade the skin, for example, and never penetrate deeper, while some proliferate and kill their host in days or hours (Siegmund, 1979). Apparently, certain viruses have mutated so that they can attack a new host, and do so vigorously. The virus of avian influenza exists in many strains; some are highly virulent—the disease is called fowl plague and is an eradicable disease—while others are quite innocuous.

The methods used to control animal diseases have evolved from the early ideas that the diseases of man and beast represented divine retribution for sins and had to be endured with patience. But in Germany during the seventeenth century, the desperate and heroic measures to control rinderpest (cattle plague), together with changes in attitude brought about by the Reformation, convinced the people that animal plagues could and must be ruthlessly stamped out. That policy was carried to England and then to the United States, where it was instituted and carried out by the Bureau of Animal Industry, USDA, in a series of brilliant programs, and the health of our animals was greatly improved (Stalheim, 1984). At the present time, surveillance is emphasized together with a readiness to cope with outbreaks of infectious diseases. On the level of an infected animal, control and cure is often accomplished by the veterinarian and the owner using an appropriate combination of testing and quarantine, chemotherapy and chemoprophylaxis, serum therapy and immunization, and other specialized procedures.

The following tabular lists of 218 animal diseases are quite limited. Many minor diseases have been omitted as well as immunologic diseases, diseases of laboratory animals, nutritional deficiencies, toxicoses, behavioral problems, and diseases due to stress. However, the diseases of fish (Snieszko, 1970), marine mammals (Fowler, 1978), and fur-bearing animals (Lybashenko, 1973) now command much attention by a variety of biologists. In most cases, the names of the diseases conform to those selected for the *Animal Diseases Thesaurus* (Veterinary Services, USDA, 1984), which was based on the 1971 *Veterinary Subject Headings* of the Commonwealth Agricultural Bureau in England. Diseases that afflict more than one species have been listed under the most common host. For a fuller description of the geographic distribution of diseases, the reader is directed to Odend'Hal (1983) and Trevino and Hyde (1984).

Table 4
Diseases of Poultry

Name	Hosts	Severity	Incidence	Distribution
Adenovirus, avian	Chickens, turkeys, quail	Mild	Unclear	Probably global
Aspergillosis	Chickens, turkeys	Mild	Sporadic	Unknown
Avian chlamydiosis	Poultry, man	Severe	Epizootic	Global
Avian encephalomyelitis	Chickens, turkeys	Severe	Unknown	Global
Avian infectious bronchitis	Chickens	Mild to severe	Diminished by vaccination	Global
Avian infectious bursal disease (Gumboro disease)	Chickens, turkeys, ducks	Severe	Common to rare	Global
Avian infectious laryngotracheitis	Chickens, peafowl	Severe	Diminished by vaccination	Global
Avian influenza	Chickens, turkeys, ducks, quail, wild and migratory birds	Mild to severe	A subtype causes fowl plague, an eradicable disease	Global
Avian leukosis	Chickens mainly	Severe	Diminished by vaccination	Global
Avian viral arthritis	Chickens mainly	Mild to severe	Unclear	Probably global
Candidiasis	Chickens, turkeys	Mild	Sporadic	Unknown
Colibacillosis	Poultry	Severe	Sporadic	Unknown
Dissecting aneurysm	Turkeys	Severe	Sporadic	North America, England
Duck hepatitis	Ducks, geese	Severe in young ducks	Common in some areas	Global
Duck plague	Ducks, geese	Severe	Unknown; not in wild ducks	North America, Europe, Asia

232

Disease	Hosts	Severity	Occurrence	Distribution
Erysipelas	Turkeys, pigs	Severe	Epizootic	Probably global
Fowl cholera	Wild and domestic birds	Very severe	Chronic to epizootic	Global
Fowl pox	All birds	Dermal only	Common	Global
Infectious coryza	Chickens	Mild to severe	Epizootic	Southern United States; probably global
Listeriosis	Chickens, ducks	Mild to severe	Sporadic	Unknown
Marek's disease	Chickens	Severe	Common; diminished by vaccination	Global
Mycoplasmosis	Chickens, turkeys	Mild to severe	Sporadic	Global
Necrotic dermatitis	Chickens	Severe	Epizootic	Unknown
Necrotic enteritis	Chickens	Severe in young chickens	Epizootic	North America
Newcastle disease	Chickens, turkeys, et al.	Mild to severe	Common; diminished by vaccination	Global
Omphalitis	Chickens, turkeys	Severe in young birds	Sporadic	Unknown
Paratyphoid	Poultry	Severe	Sporadic	Unknown
Pasteurellosis (new duck disease)	Ducks, turkeys, chickens	Mild to serious	Sporadic	Unknown
Salmonellosis	Chickens, turkeys	Severe	Diminished by testing	Global
Spirochetosis	All poultry	Mild to severe	Sporadic	Global
Transmissible turkey enteritis	Turkeys	Severe in young birds	Unknown	Unknown
Tuberculosis	Poultry, pigs	Severe	Sporadic	Global
Ulcerative enteritis	Quail	Very severe	Epizoonotic	Unknown
Vibrionic hepatitis	Chickens	Mild to severe	Sporadic	North and South America, Europe

Table 5
Diseases of Sheep and Goats

Name	Hosts	Severity	Incidence	Distribution
Bluetongue	Sheep, goats, cattle, deer	Mild to severe	Not in northeastern United States or Canada	South Africa, Australia, parts of United States
Border disease	Sheep	Severe in lambs	Not known	New Zealand and probably global
Borna disease	Sheep, horses	Mild to severe	Not outside Europe	Central Europe
Brucellosis	Goats, man	Mild	Much diminished	Southwestern United States; Middle East
Caseous lymphadenitis	Sheep, horses	Chronic	Sporadic	Global
Clostridial disease	Sheep, goats	Severe	Sporadic	Global
Contagious caprine pleuro-pneumonia	Goats	Mild to severe	Sporadic, epizootic	Mexico, United States, Middle East, Africa
Contagious Ecythema (sore mouth)	Sheep, goats	Mild to severe	Common	Global
Endemic abortion	Sheep	Mild	Sporadic	North America, Europe
Erysipelas	Sheep	Severe	Sporadic	Unknown
Foot rot	Sheep, cattle	Mild to severe	Sporadic	Global

234

	Ruminants			
Heartwater		Mild to severe	Sporadic, epizootic; not in United States	Africa, Europe
Louping ill	Sheep mainly	Mild to severe	Not in United States	British Isles
Nairobi sheep disease	Sheep, goats	Mild	Not in United States	Africa
Ovine genital campylobacteriosis	Sheep	Mild	Sporadic, epizootic	Probably global
Ovine progressive pneumonia (Maedia-Visna)	Sheep, goats	Severe	Enzootic	North America, Europe, Asia
Pasteurellosis	Sheep	Severe	Enzootic	Unknown
Peste des petits	Ruminants, sheep, goats	Severe	Not in United States	Africa
Preganancy toxemia	Ewes	Severe	Sporadic	Global
Pulmonary adenomatosis	Sheep	Severe	Unknown	Global, except Australia
Q fever	Sheep, cattle, man	Mild	Epizootic	Unknown
Rift Valley fever	Sheep, goats, cattle, buffalo, man	Severe	Not in United States	Africa
Scrapie	Sheep, goats	Severe	Sporadic	Probably global
Sheep pox	Sheep	Very severe	Not in United States	Africa, Asia
Tetanus	Sheep	Very severe	Sporadic	Global
Tularemia	Sheep	Mild to severe	Sporadic	Western United States
Wesselsbron virus	Sheep, cattle	Mild	Not in United States	Africa

235

Table 6
Diseases of Dogs

Name	Hosts	Severity	Incidence	Distribution
Brucellosis	Dogs	Mild, venereal	Diminished by testing	Unknown
Canine coronavirus	Dogs, coyotes	Mild to severe	Common; newly recognized	North America, Europe, Australia
Canine distemper	Dogs, coyotes, foxes, et al.	Mild to severe	Diminished by vaccination	Global
Canine herpes	Dogs	Severe in puppies	Probably common	Global
Canine infectious hepatitis	Dogs, foxes, coyotes	Inapparent to severe	Diminished by vaccination	All developed countries
Canine parvovirus	Dogs	Severe	Common	A new global pathogen
Leptospirosis	Dogs, cattle, man	Mild to severe	Common	Global
Lyme disease	Dogs	Mild to severe	Newly recognized	North America, Europe
Rabies	Most warm-blooded animals	Severe	Sporadic; diminished by vaccination	Global, except British Isles, New Zealand
Tuberculosis	Dogs, cattle, man	Mild	Much diminished	Unknown

Table 7
Diseases of Cats

Name	Hosts	Severity	Incidence	Distribution
Feline calcivirus	Felidae	Mild to severe	Diminished by vaccination	North America, Europe, Australia
Feline cytauxzoonosis	Cats	Severe	Sporadic	Southern United States
Feline infectious peritonitis	Felidae	Usually mild	Unknown	Probably global
Feline leukemia	Felidae	Severe	Common	Probably global
Feline panleukemia	Felidae, mink	Severe in young cats	Diminished by vaccination	Global
Feline rhinotracheitis	Felidae	Mild	Unknown	Probably global
Plague	Cats, man	Mild to severe	Rare	Western United States

Table 8
Diseases of Fish

Name	Severity	Incidence	Distribution
Bacterial cold water disease	Mild; severe in young fish in cold water	Sporadic	Mainly the Pacific Northwest
Bacterial gill disease	Mild and chronic	Endemic	Mainly in northwestern United States
Carp pox	Mild	Sporadic	North America, Japan
Channel catfish virus disease	Severe in young fish	Sporadic	North and South America
Columnaris disease	Mild; severe in warm waters	Sporadic	Probably all fish in United States
Furunculosis	Mild to severe when stressed	Sporadic	Common in United States
Herpes virus disease	Severe	Newly recognized in salmonids, turbot	Unknown
Ichthyophonus	Mild	Epizootic	Marine fish
Infectious hemopoietic necrosis	Severe in young fish	Sporadic	North America, Japan
Infectious pancreatic necrosis	Mild to severe	Common	North America, Europe, Japan
Lymphocystis disease	Chronic	Low to high	Marine and freshwater fish
Pseudotuberculosis	Severe when stressed	Sporadic	Japan, United States
Sporocytophaga disease	Mild to severe	Unknown	Marine fish
Ulcerative dermal necrosis	Salmon	Sporadic	Unknown
Vibriosis	Severe when stressed	Diminished by vaccination	Marine fishes

Table 9

Diseases of Marine Mammals—Pinnipeds and Cetaceans

Name	Hosts	Severity	Incidence	Distribution
Aspergillosis	Pinnipeds	Chronic	Unknown	Probably global
Candidiasis	Cetaceans	Severe, chronic	Common	Unknown
Clostridial myositis	Cetaceans	Severe	Rare	Unknown
Coccidioidomycosis	Sea lions	Severe	Rare	Unknown
Cutaneous strepto-trichosis	Pinnipeds	Chronic	Unknown	Unknown
Erysipelas	Captive animals	Severe	Enzootic	Unknown
Leptospirosis	Wild pinnipeds	Severe	Epizootic	California coastal waters
Nocardiosis	Cetaceans	Severe	Sporadic	Unknown
Seal pox	Cetaceans	Mild	Sporadic	Unknown

Table 10
Diseases of Fur-Bearing Animals

Name	Hosts	Severity	Incidence	Distribution
Aleutian disease	Mink	Severe; diminished by testing	Low	Probably global
Aujesky's disease	Mink	Severe	Sporadic, from pork	Unknown
Botulism	Mink	Very severe	Sporadic; reduced by vaccination	Unknown
Distemper	Mink (dogs)	Severe	Sporadic; diminished by vaccination	Global
Hemorrhagic pneumonia	Mink	Severe	Sporadic	Unknown
Infectious myxomatosis	Rabbits	Very severe	Epizootic	Western United States, Europe, Australia
Listeriosis	Chinchilla, rabbits	Severe	Sporadic	Unknown
Mink viral enteritis	Mink	Severe in young mink	Sporadic	North America
Necrobacillosis	Rabbits	Chronic	Sporadic	Unknown
Papillomatosis	Rabbits	Chronic	Sporadic	Unknown
Pasteurellosis	Rabbits	Mild; snuffles, abscesses, or conjunctivitis	Common	Unknown
Rabbit pox	Rabbits	Severe	Not in wild rabbits	Unknown
Ringworm	Rabbits, chinchilla	Mild	Sporadic	Probably global
Toxoplasmosis	Chinchilla	Severe	Sporadic	Unknown
Transmissible mink encephalopathy	Mink	Very severe	Sporadic	Unknown
Treponematosis	Rabbits	Mild to severe	Sporadic, venereal	Unknown
Tuberculosis	Mink, rabbits	Mild to severe	Enzootic	Probably global
Yersiniosis	Rabbits	Chronic	Sporadic	Unknown

References

Fowler, M. E., ed. (1978). "Zoo and Wild Animal Medicine," pp. 587–589. W. B. Saunders Co., Philadelphia.

Hayes, J., ed. (1984). "Animal Health. Yearbook of Agriculture." U.S. Government Printing Office, Washington, D.C.

Lybashenko, S. Y., ed. (1973). "Diseases of Fur-Bearing Animals," pp. 39–209. Oxonian Press, New Delhi.

Odend'Hal, S. (1983). "The Geographical Distribution of Animal Viral Diseases." Academic Press, New York.

Siegmund, O. H., ed. (1979). "The Merck Veterinary Manual." Merck and Co., Rahway, New Jersey.

Snieszko, S. F., ed. (1970). "A Symposium on Diseases of Fishes and Shellfishes," pp. 231–349. American Fisheries Society, Washington.

Stalheim, O. H. V. (1984). Contributions of the Bureau of Animal Industry to the veterinary profession. *J.A.V.M.A.* **184,** 1222–1224.

Trevino, S. T., and Hyde, J. L. (1984). "Foreign Animal Diseases: Their Prevention, Diagnosis, and Control." United States Animal Health Association, Richmond.

U.S. Public Health Service (1984). "Morbidity Mortality Annual Summary," p. 3. Centers for Disease Control, Atlanta.

Production

Reproduction

H. W. Hawk

U.S. Department of Agriculture
Agricultural Research Service
Beltsville Agricultural Research Center
Reproduction Laboratory
Beltsville, Maryland

Handbook of Animal Science

Introduction

Successful animal farming depends upon obtaining the maximum number of viable offspring per year from each reproducing female. In the case of cattle, generally giving birth to one offspring per pregnancy, efficient reproduction means obtaining a calf each year from as many cows as possible; with sheep, obtaining more than one lambing per year if possible; with swine, two litters per year; with poultry, the highest possible number of eggs per year. For high reproductive efficiency, farm animals should reach puberty at an early age, possess high fertility, and raise a high proportion of their offspring.

Characteristics of Reproduction

Female Reproduction

Puberty

Puberty is the stage of development at which the animal acquires the desire and ability to mate and produces and releases functional gametes. The age of puberty (Table 1) varies greatly in the developing young of each class of animal. Variation is caused by both genetics (e.g., relatively late puberty in Zebu cattle compared to European breeds of cattle) and by environment, particularly the level of nutrition and the consequent rate of growth. Fast growth generally favors early puberty. In some species generally characterized by seasonal breeding, such as sheep, goats, and horses, puberty can be delayed for months when growing young reach breeding size during seasonal anestrus.

Reproductive Cycles

Estrous Cycles   Cattle, sheep, swine, and horses are all polyestrous, with cattle, swine, and some breeds of sheep having fertile estrous cycles all year. Most sheep, goats, and horses tend to be seasonally polyestrous, with sheep and goats having fertile estrous cycles through the autumn and winter in the Northern Hemisphere and horses having fertile cycles through the spring and summer. Thus, sheep and goats

Table 1
Characteristics of Estrus and Estrous Cycles[a]

	Cattle		Sheep	Goats	Swine	Horses
	Dairy	Beef				
Average age at puberty, males (days)	275 (225–325)	300 (250–350)	150 (120–185)	150 (120–180)	200 (150–240)	455 (300–730)
Average age at puberty, females (days)	330 (300–360)	390 (320–460)	200 (140–240)	180 (120–210)	210 (115–250)	540 (300–730)
Type of estrous cycling	Polyestrous	Polyestrous	Seasonal polyestrous	Seasonal polyestrous	Polyestrous	Seasonal polyestrous
Average length of estrous cycle (days)	21 (17–24)	21 (17–24)	16.7 (14–19)	20.6 (18–23)	21 (19–23)	21.7 (18–24)
Average duration of estrus (hr)	15 (12–20)	20 (12–30)	30 (18–40)	34 (26–42)	56 (40–70)	6.5 days (2–11 days)
Time of ovulation after beginning of estrus	28 hr	30 hr	24 hr	36 hr	40 hr	5 days
Usually number of follicles ovulated	1	1	1–6	1–4	13–22	1

[a]Compiled primarily from Cole and Cupps, 1977 (chaps. 16, 17, 18); Ginther, 1979 (chaps. 8, 9); Hafez, 1980 (chaps. 16, 17, 18).

begin estrous cycles with decreasing day length, whereas horses begin to cycle with increasing day length.

Each species has a generally recognized modal estrous cycle length (Table 1). There is considerable variation around the modes. In instances where estrous periods are either not expressed by the animal or are not detected by humans, two or more ovulation cycles will be considered as one estrous cycle, which increases and distorts the arithmetic mean length of the estrous cycle and makes the modal length more useful. Estrous cycles can be abnormally short or long. Defective formation of the corpus luteum can result in its early regression and cause short estrous cycles. The first estrous cycle after parturition in cattle is often short. Long estrous cycles can result not only from the addition of two or more ovulation cycles but from prolongation of corpus luteum function, often by a pregnancy that was lost after the presence of an embryo had prolonged the estrous cycle. Cycles can also be lengthened by uterine infection or inflammation that interferes with the physiological mechanisms in the uterus that initiate regression of the corpus luteum.

Estrus and Ovulation   The duration of estrus varies considerably among estrous periods (Table 1). In modern farming, the duration of estrus in dairy cattle is particularly important because a high proportion of dairy females are inseminated artificially and estrus must be detected by humans. The physiological bases of variation in the duration and intensity of estrus are not well understood, but environmental effects can be important. For example, high temperature during summer often suppresses the expression of estrus in cattle. Slippery floors will seriously inhibit the desire of cows to mount or to stand for mounting by other cows (the most obvious signs of estrus). Short estrous periods, particularly of low intensity, are often not detected by humans because estrus may fall between periods of observation, particularly at night. The system of management of dairy cattle and the effort devoted to estrus detection by dairymen often determines the efficiency of estrus detection; only about half of expected estrous periods are actually detected.

Ovulation generally occurs near the end of estrus (Table 1), although cattle ovulate about 12 hr after the end of estrus. Cattle occasionally ovulate more than one follicle, and the frequency of twinning is about 1.2%. In sheep and goats, the number of ovulations depends upon such factors as breed, age, and stage of the breeding season. In mares, the frequency of twin ovulations is about 16%, but embryonic and fetal mortality is high and the frequency of twinning is only 1 to 2%.

Fertilization, Implantation, and Gestation   Ova in cattle, swine, and horses must be fertilized within a few hours; ova in goats and sheep live

Table 2

Embryo Development and Transfer[a]

	Cattle	Sheep	Goats	Swine	Horses
Fertile life of ova (hr)	8–20	16–24	12–24	8–10	6–8
Time of development to:					
2-cell	1 day	1 day	1.5 days	14–20 hr	1 day
8-cell	3 days	2.5 days	3 days	2.5 days	3 days
Blastocyst	7–8 days	6–7 days	5–6 days	5–6 days	6 days
Oviduct transit (days)	3.5–4.5	2.5–3.75	2–4	2.5–3.75	5–5.5
Time of implantation (days)	34–40	15–17	15–20	11–20	49–63
Average length of gestation (days)	281 (268–295)	148 (138–159)	149 (138–159)	114 (110–120)	338 (330–345)
Type of placenta	Cotyledonary	Cotyledonary	Cotyledonary	Diffuse	Diffuse

[a]Compiled primarily from Ginther, 1979 (chaps. 8, 9); Hafez, 1980 (chaps. 12, 13, 29); Winnsatt, 1975.

Table 3
Reproductive Function after Parturition[a]

	Cattle		Sheep	Swine	Horses
	Dairy	Beef			
Parturition to:					
Ovulation (days)	20	62	18[b]	9[c]	12
	(10–50)	(35–105)		(7–12)	(9–16)
Estrus (days)	34	63	35[b]	7[c]	8
	(20–70)	(40–110)		(4–9)	(5–12)
Uterine involution	45	45	27	24	20
(days)	(32–50)	(32–50)	(20–35)	(18–30)	(10–40)

[a]Data compiled primarily from Ginther, 1979 (chap. 11); Hafez, 1980 (chaps. 16, 17, 18). [b]After parturition during the breeding season. The first ovulation after parturition in sheep and dairy cattle is seldom accompanied by estrus. Ovulation and estrus may be delayed by months after parturition during seasonal anestrus in sheep and horses. [c]Days postweaning.

somewhat longer before they must be fertilized (Table 2). Ova are fertilized in the ampulla, the anterior segment of the oviduct, and fertilized ova generally reach the uterus by the fourth day after ovulation. Embryonic membranes attach to the uterine endometrium by about 34 days in cattle, 16 days in sheep and goats, and 55 days in horses. The attachments are initially loose, gradually becoming more firm. The length of gestation varies somewhat within each species, and even breeds of cattle differ in mean length of gestation. In dairy breeds, Ayrshires average 279 days and Brown Swiss 290 days; in beef breeds, Angus average 279 days and Herefords 285 days. The placenta is cotyledonary except in swine and horses, which have a diffuse type of placental attachment (Table 2).

Resumption of Estrous Cycles after Parturition   After parturition, the uterus gradually returns toward its original size, a process termed involution, although the uterus usually remains larger after the birth of the first offspring than it was before the first pregnancy. Characteristic time to involution for each species is given in Table 3. Ovulation resumes within a few weeks after parturition in dairy cattle and in sheep, goats, and horses if parturition occurs during the breeding season (Table 3). Estrus and ovulation tend to be delayed during lactation, often termed lactational anestrus, in beef cattle, sheep, and swine. Delays in resumption of estrous cycling and ovulation can be caused by inadequate nutrition in beef cattle and sometimes in sheep. Swine seldom exhibit estrus in association with ovulation until after their young are weaned (Table 3).

Table 4
Hormones Controlling Reproduction[a]

Hormone	Source	Sex affected	Function
Estrogen	Ovarian follicles	Females	Induces estrous behavior; stimulates growth and function of reproductive tract; stimulates uterine contractions; regulates gonadotropin release; develops and maintains secondary sexual characteristics. In poultry, stimulates growth of oviduct and stimulates uterine secretions to calcify egg shell.
Progesterone	Ovarian corpus luteum (largest follicles in poultry)	Females	Prepares uterus for implantation; stimulates endometrial secretions; maintains pregnancy; suppresses LH surges; generally antagonizes estrogenic effects (in small amounts, progesterone synergizes with estrogen to enhance estrogenic effects on estrous behavior). In poultry, triggers LH release by hypothalamus, development of oviduct.
Follicle-stimulating hormone (FSH)	Anterior pituitary	Females and males	Stimulates follicle growth and estrogen secretion in females; spermatogenesis in males.
Luteinizing hormone (LH)	Anterior piuitary	Females and males	Causes ovulation of mature follicles; stimulates progesterone and estrogen secretion by corpus luteum and follicles; stimulates androgen secretion in males.
Prolactin	Anterior pituitary	Males and females	May stimulate corpus luteum function and progesterone secretion in some species.
Gonadotropin-releasing hormone	Hypothalamus	Females and males	Causes release of FSH and LH from anterior pituitary.
Prostaglandin F_2	Uterus	Females (except poultry)	Probably the natural hormone that causes regression of corpus luteum in sheep, cattle, and swine.

(*continued*)

Table 4 (*Continued*)

Hormone	Source	Sex Affected	Function
Androgens	Testes	Males	Stimulates spermatogenesis, sexual behavior, and development and maintenance of accessory sex glands and secondary sexual characteristics.

[a]Information obtained primarily from Ginther, 1979 (chap. 5).

Reproductive Hormones

Hormones are chemical mediators produced by endocrine glands that regulate the function of organs in another part of the body. Reproductive hormones are produced by the hypothalamus, pituitary gland, ovaries, testes, and placenta (Table 4). Hormones regulate all of the critical steps in successful reproduction in both sexes, including puberty, ovulation, transport of sperm and ova in the female reproductive tract, maintenance of pregnancy, and parturition. In the male, hormones regulate puberty and sperm production, as well as development and maintenance of accessory sex glands and secondary sexual characteristics.

Two hormones can act in either a complementary or antagonistic manner; for example, a small amount of estrogen or progesterone generally facilitates the actions of the other, whereas a larger amount of one hormone antagonizes the actions of the other. Much has been learned in recent years about the amounts of hormones secreted and their actions by the development and use of radioimmunoassays and radioreceptor assays. Relative changes in hormones found in the circulation during the estrous cycle are given in Fig. 1.

Fertility

Fertility of domestic animals ranges from as high as 85% in sheep and goats mated naturally in the middle of the breeding season to as low as 50–55% or lower in dairy cattle inseminated artificially. The level of fertility depends upon such factors as the time and method of inseminations, fertility of the sire, and normality of reproductive function of the female. High environmental temperatures depress fertility of both males and females. Most common causes of infertile services are death of embryos developing in the uterus, most of which occurs within the span

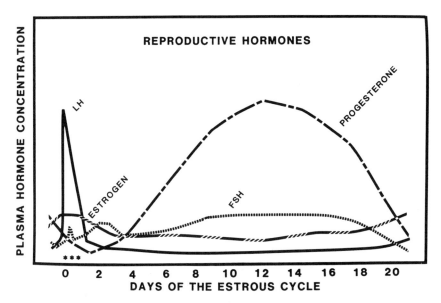

Figure 1   Relative changes in reproductive hormones throughout the estrous cycle in heifers or gilts. Graph begins at estrus (***). Hormonal changes in ewes are similar except that the cycle length is about 4 days shorter.

of a normal estrous cycle, fertilization failure of ova in animals inseminated at the proper time, and, in artificially inseminated cattle, improperly timed insemination (Table 5). In cattle inseminated artificially at the proper time, about 13% of normal ova are not fertilized; in naturally

Table 5

Major Causes of Reproductive Losses after Insemination[a]

Type of Loss	Cattle[b]	Sheep	Swine	Chickens	Turkeys[b]
Improperly timed insemination (%)	10	—	—	—	—
Ovulation failure (%)	2	1	1	—[c]	—[c]
Lost or ruptured ova (%)	5	5	1	10	10
Fertilization failure (%)	13	12	5	5	10
Embryonic death (%)	18	17	28	10	20
Stillbirths and neonatal deaths (%)	6	10	10	—	—

[a]Primarily from Hafez, 1980 (chaps. 16, 17, 18); Hawk, 1979 (chaps. 1, 2, 3, 4, 5); Quinlaven et al., 1966. [b]Artificially inseminated. [c]Failure to lay an egg daily could be considered to be ovulation failure in poultry.

mated cattle, the rate of fertilization failure is substantially less, probably mostly due to the deposition of more sperm in the female by natural mating than by artificial insemination. Stillbirths and neonatal deaths, averaging 5 to 10%, are particularly wasteful to animal producers because the female has been maintained throughout pregnancy.

Male Reproduction

Semen and Sperm

Ages at puberty for males of various species were listed in Table 1. Characteristics of semen and sperm from males are given in Table 6. At natural mating, all males except swine and horses deposit semen in the vagina; boars and stallions deposit semen in the uterus.

Manipulation of Reproduction

Artificial Insemination

General Aspects

Artificial insemination has been practiced in farm animals for several decades. Approximately 60% of dairy cattle in the United States but only about 5% of beef cattle and only small numbers of sheep, goats, and swine are inseminated artificially. All turkeys are inseminated artificially because the heavy breasts of tom turkeys prevent them from mounting females. Frequent handling of dairy cattle by humans facilitates observations for estrus and use of artificial insemination; the availability and high fertility of frozen bull semen are also important. Infrequent handling of beef cattle and sheep generally limits the use of artificial insemination in these animals. Several factors have delayed the adoption of artificial insemination in swine, including the labor required for collecting boar semen on farms, the limited number of inseminations per ejaculate, and the lower fertility of frozen semen. Technological developments during the last few years, particularly the ability to freeze and thaw semen and obtain reasonably high fertility and to synchronize estrus in females, make artificial insemination of beef cattle and sheep more feasible. However, fertility may suffer somewhat when frozen semen is used for artificial insemination, especially with sheep and swine. The composition of several semen extenders commonly used for artificial insemination are given in Table 7. Other aspects of artificial insemination for the various animals are given in Table 8.

Table 6
Characteristics of Semen and Sperm[a]

Characteristic	Cattle		Sheep	Goats	Swine[b]	Horses[b]	Poultry	
	Dairy	Beef					Chickens	Turkeys
Volume (ml)	7 (4–10)	5 (3–8)	1 (0.7–1.4)	1 (0.5–1.5)	225 (150–300)	60 (30–100)	.6 (0.4–1.0)	.3 (0.1–0.6)
Sperm concentration (10^9/ml)	1.2 (0.5–2)	1.0 (0.4–1.5)	3 (1–5)	3 (2–6)	.2 (0.1–0.3)	.2 (0.15–0.3)	5 (3–8)	10 (5–15)
Total sperm (10^9)	8 (5–15)	5 (4–8)	3 (1.5–4.0)	2 (1.5–5.0)	45 (30–60)	9 (5–15)	3 (3–8)	3 (2–4)
Motile sperm (%)	70 (50–80)	65 (40–75)	70 (60–80)	70 (60–80)	60 (50–80)	60 (40–75)	85 (70–90)	60 (45–70)
Morphologically normal sperm (%)	80 (70–95)	75 (65–90)	90 (80–95)	90 (80–95)	70 (60–90)	70 (60–90)	90 (80–95)	85 (75–90)
Ejaculates per week	4 (2–6)	4 (2–6)	20 (7–25)	15 (7–20)	3 (2–5)	3 (2–6)	3 (2–5)	3 (2–4)

[a]Compiled primarily from Cole and Cupps, 1977 (chaps. 9, 10); Hafez, 1980 (chap. 26). [b]Gel-free portion of ejaculate.

Table 7
Semen Extenders for Artificial Insemination with Frozen Semen[a]

Ingredients	Bull[b]	Bull[b]	Boar[b]	Ram	Buck (Goat)	Stallion	Turkey[b]
Tris (hydroxymethyl) Aminomethane (g)	24.20	—	2.00	36.34	24.20	—	—
Tes [N-Tris (Hydroxymethyl)] Methyl-2-Aminoethanesulfonic Acid (g)	—	—	12.00	—	—	—	1.95
Dipotassium phosphate trihydrate (g)	—	—	—	—	—	—	12.70
Monopotassium phosphate (g)	—	—	—	—	—	—	0.65
Magnesium chloride sextahydrate (g)	—	—	—	—	—	—	0.34
Sodium acetate trihydrate (g)	—	—	—	—	—	—	4.30
Sodium glutamate (g)	—	—	—	—	—	—	8.60
Glucose (g)	10.00	—	32.00	5.00	10.00	50.00	—
Fructose (g)	—	—	—	5.00	—	—	5.00
Citric acid monohydrate (g)	13.40	—	—	—	13.40	—	—
Potassium citrate (g)	—	—	—	—	—	—	0.64
Lactose (g)	—	—	—	—	—	3.00	—
Raffinose (g)	—	—	—	—	—	3.00	—
Orvus ES paste (g)	—	—	5.00	—	—	—	—
Penicillin (units/ml)	1000	1000	—	—	1000	1000	—
Streptomycin (g/ml)	1000	1000	—	—	1000	1000	—
Polymyxin B (units/ml)	500	500	—	—	—	—	—
Cow milk (ml)	—	930[c]	—	—	—	—	—
Egg yolk (ml)	200	—	200	150	100	50	—
Glycerol (ml)	70	70	10	50	80	50	—
Dimethylsulfoxide	—	—	—	—	—	—	40
Distilled water to final volume (ml)	1000	1000	1000	1000	1000	1000	1000

[a]Primarily from Hafez, 1980 (chap. 26); Pursel and Johnson, 1975; Salamon, 1976; Sexton, 1980. [b]Used for insemination of fresh (unfrozen) bull semen with deletion of glycerol and for insemination of turkey semen with deletion of dimethylsulfoxide. Pursel and Johnson, 1975; J. Anim. Sci. **40**, 99–107. [c]Whole or skimmed milk, heated to 95° F for 15 min.

Table 8
Artificial Insemination of Domestic Animals with Frozen Semen[a]

	Cattle	Sheep	Goats	Swine	Horses	Turkeys
Volume of inseminate (ml)	0.2–1	0.05–0.2	0.02–0.2	50	20–50	0.025–0.05
Number of motile sperm per inseminate (10^6)	15	120–150	100–150	5000	1500	100
Time of insemination	8–16 hr after onset of estrus	10–20 h after onset of estrus	12–36 h after onset of estrus	15–30 h after onset of estrus	Every 2 days during estrus	Weekly
Site of semen deposition	Uterus–cervix	Cervix	Cervix or uterus	Uterus	Uterus	Vagina

[a] Primarily from Cole and Cupps, 1977 (chap. 10); Ginther, 1979 (chap. 8); Hafez, 1980 (chaps. 17, 26).

Semen Collection and Insemination Procedures

Semen is collected from males (except poultry) by use of an artificial vagina. Semen is collected from turkeys and chickens manually by abdominal massage.

Cattle and horses are artificially inseminated by inserting a cannula through or into the cervix and depositing semen in the body of the uterus. In some heifers and in most sheep, the cannula cannot be passed through the cervix, so semen is deposited within the cervix. In swine, a specially designed insemination tube is threaded into the cervix and semen is expelled into the uterus. To inseminate poultry, the oviduct is everted using slight manual pressure, and semen is deposited into the vagina.

At the present time, almost all bull semen used for artificial insemination is frozen in straws. The semen can be stored for months or years and used for planned matings long after a bull is no longer in service.

Regulation of Estrus and Ovulation

The actual purpose of regulating estrus is to control the time of ovulation. Estrus can be regulated and generally timed within 1 to 2 days in farm animals. Animals may be inseminated without detection of estrus,

Table 9
Classes of Compounds Used for Synchronization
of Estrus in Farm Animals[a]

Class of compound	Route and length of administration	Class of animal
Progestogen	Oral, intravaginal pessary, injection, subcutaneous implant; 12–21 days (approximately one estrous cycle)	Cattle, sheep, goat, swine (generally effective in acyclic as well as cycling females)
Prostaglandin F_2 or analog	Intramuscular injection, once during luteal phase of estrous cycle	Cattle, sheep, goat, horse
Progestogen with estrogen	Progestogen as above, 9–12 days; estrogen by intramuscular injection at beginning of treatment to shorten life of corpus luteum	Cattle
Progestogen and prostglandin F_2	Progestogen by intravaginal pessary, 5–7 days; prostaglandin near last day to regress corpus luteum	Cattle

[a]Primarily from Hafez, 1980 (chap. 27); Hawk, 1979 (chap. 8).

although fertility is reduced. Two classes of compounds, progestogens and prostaglandin F_2 (Table 9), can be used to regulate the time of estrus in animals undergoing ovulatory cycles. Progestogen can be administered for approximately the length of an estrous cycle to artificially prolong the luteal phase of the cycle. If estrogen is given at the beginning of treatment to cause the early regression of corpora lutea or inhibit the development of corpora lutea, progestogen treatment can be shortened. Within 2 to 5 days after withdrawal of progestogen, a high proportion of females will be in estrus. Fertility at the synchronized estrus is often lower than normal. Progestogen can also be used to induce estrus in anestrous or prepuberal females after the treatment is withdrawn.

Prostaglandin F_2 or its analogs, when administered by intramuscular injection, cause premature and rapid regression of corpora lutea. Most females will be in estrus in 2 to 4 days. Prostaglandin is not effective in females without corpora lutea, either before puberty, during anestrus, or during the follicular phase of the estrous cycle. However, if prostaglandin is administered twice, 8 to 12 days apart, almost all animals will be in synchronized estrus after the second injection. At the time of the second injection, almost all animals should have a corpus luteum of the proper age to be regressed by the treatment.

Superovulation and Embryo Transfer

Farm animals can be treated with hormones to cause them to superovulate (i.e., to ovulate more ova than normal). In dairy cattle, the number of ova shed at superovulation often averages 20 or more instead of the one ovum normally shed.

Superovulation involves the administration of a gonadotropin to stimulate the growth of more ovarian follicles than would normally develop. The gonadotropin is usually follicle-stimulating hormone (FSH) or pregnant mare serum gonadotropin (PMSG).

In cattle, the gonadotropin treatment is usually begun near the middle of the estrous cycle. FSH is given by injection twice daily for 4 or 5 days. PMSG, which is more resistant to degradation in the injected animal than is FSH, is given only once. About 3 days after gonadotropin treatment has begun, prostaglandin F_2 or an analog is injected to regress the corpus luteum. The animal is usually in estrus 2 days after prostaglandin treatment. Cows are usually inseminated two or three times in attempts to maximize the fertilization rate, which averages about 65% in superovulated cattle. The ova enter the uterus from the oviducts 3 to 4 days after estrus. Ova can be collected from a superovulated animal at any time after fertilization, but collection before 4 or 5 days after estrus requires surgery. In the practice of embryo transfer with cattle, ova are

Table 10
Current Estimates of Reproductive
Efficiency in Farm Animals

Animal	Standard of Performance[a]	Estimate of Current Performance[b]	Attainable Performance
Dairy cattle	Calving interval	13.5–14 mo	12.5 mo
Beef cattle	Calf crop/cow/yr	75–79%	90%
Sheep	Lambs weaned/ewe/yr	1.5	3.0
Swine	Pigs marketed/sow/yr	13	20
Chickens	Eggs laid/yr	220	250
Turkeys	Eggs laid/yr	95	125

[a]Commonly used standards of reproductive performance. [b]Data obtained primarily from Hawk, 1979 (chaps. 1, 2, 3, 4, 5).

usually collected nonsurgically by flushing the uterus through a catheter about 7 days after estrus. Embryos are generally in the blastocyst stage of development at that time, and they can be transferred either surgically or nonsurgically into recipient cattle. The conception rate after embryo transfer is usually 60–70%, with the conception rate after surgical transfer being slightly higher than after nonsurgical transfer. Embryos are sometimes split into equal halves, and two calves can often be obtained by transferring the halves into one or two recipient animals.

Reproduction Efficiency

At the present time, the reproductive efficiency of farm animals is considerably below that which could be obtained by the application of management practices and technological developments that would improve reproductive performance. Practices that can be widely applied range from efficient detection of estrus in artificially inseminated dairy cattle to synchronization of estrus and use of artificial insemination in beef cattle and sheep. Estimates of current reproductive efficiency and reasonably attainable levels of efficiency are given in Table 10.

References

Cole, H. H., and Cupps, P. T., eds. (1977). "Reproduction in Domestic Animals." Academic Press, New York.
Ginther, O. J. (1979). "Reproductive Biology of the Mare." McNaughton and Gunn, Inc., Ann Arbor, Michigan.

Hafez, E. S. E., ed. (1980). "Reproduction in Farm Animals," 4th ed. Lea and Febiger, Philadelphia, Pennsylvania.

Hawk, H. W., ed. (1979). "Beltsville Symposia in Agricultural Research 3: Animal Reproduction." Allanheld, Osmun, Montclair, New Jersey.

Pursel, V. G., and Johnson, L. A. (1975). Freezing of boar spermatozoa: Fertilizing capacity with concentrated semen and a new thawing procedure. *J. Animal Sci.* **40,** 99.

Quinlivan, T. D., Martin, C. A., Taylor, W. B., and Cairney, I. M. (1966). Estimates of pre- and perinatal mortality in the New Zealand Romney Marsh ewe. *J. Reprod. Fert.* **11,** 379.

Salamon, S. (1976). "Artificial Insemination of Sheep." Publicity Press, Chippendale, NSW, Australia.

Sexton, T. J. (1980). A new poultry semen extender 5. Relationship of diluent components to cytotoxic effects of dimethylsulfoxide on turkey spermatozoa. *Poultry Sci.* **59,** 1142.

Winnsatt, W. A. (1975). Some comparative aspects of implantation. *Biol. Reprod.* **12,** 1.

Growth Rates

Theron S. Rumsey

U.S. Department of Agriculture
Agricultural Research Service
Beltsville, Maryland

Table 1
Average Weight Gains of Domestic Animal
Species during Growth and Finishing

Species	Growth rate (g/day)		Reference
	Male	Female	
Rabbit	39	—	Cheeke, 1983
Chicken (broiler)	25	22	Thomas and Bossard, 1982
Turkey (tom)	85	—	Moran et al., 1973
Pig	846	776	Bereskin, 1983
Lamb	302	241	Stritzke and Whiteman, 1982
Beef			
Medium frame	1200	1000	National Research Council, 1984
Large frame	1600	1200	Rand et al., 1958

Introduction

Growth rate of food-producing animals is the primary measure used by the producer as a quality indicator of his animals' performance, management system effectiveness, and profitability. Growth rates vary widely depending on animal species. Under optimum conditions, growth rate for food-producing animals varies from less than 30 g per day for broilers to more than 1500 g per day for feedlot steers (Table 1). Growth rate also varies within a species depending on diet, genetic potential, health, and age. To the animal producer, who knows the optimum rates achievable with his particular production system, growth is recognized as the overall end result of many factors and an indicator of the health and well-being of the animals in that system. If a part of the system is changed or if a health issue arises, these usually affect growth rate. Growth rate is an important monitor in animal nutrition research, as single factors are changed with the objective of improving production efficiency. Ultimately, growth rate of animals in a particular animal production system translates into profitability, as greater growth rates dilute maintenance costs and thus improve efficiency. Profitability of animal agriculture is based on a producer's ability to market feedstuffs, grain and forage, through conversion to animal foods. If animal production is insufficient to pay the costs of the feed and management for those animals, the system is not profitable. Also, in large intensive production systems, facility costs and loan interests are often large expenses; thus, maximizing growth rate often is the predictor of profit or loss through savings of fixed daily costs.

Diet

The type of feedstuff or the type of diet that an animal eats and the intake of that diet are important factors in determining growth rate. Under conditions where diets are formulated to meet all known nutrient requirements, the energy density of the diet is a regulator of feed intake and of growth rate (Table 2). In general, the more fiber or lower digestibility of the diet the less growth per unit of diet consumed, thus growth rate becomes a function of feed intake and quality of the diets. In monogastrics, this is true up to a point where energy intake equals the animal's ability to utilize or deposit that energy. Then intake will tend to decrease slightly; for example, when fat is added to the diet, to give a further improvement in feed conversion. To further increase growth rates in monogastrics under optimum nutritional status requires changes in the animal's genetic or physiological ability to deposit tissue. Ruminant animals are similar to monogastric animals in their use of energy-dense diets in that an upper limit to energy intake is achieved that is based on genetic and physiological limits. Unlike the monogas-

Table 2
Feed Intakes and Growth Rates of Animals
Fed Diets of Different Energy Densities

Diet	Animal	Intake (g/day)	Gain (g/day)	Reference
Corn concentrate				
−Fat	Broiler	22	11	Rand *et al.*, 1958
+Fat	Broiler	17	13	Rand *et al.*, 1958
Corn concentrate				
−Fat	Pig	1462	424	Mersmann *et al.*, 1984
+Fat	Pig	1284	466	Mersmann *et al.*, 1984
Grain concentrate				
Corn	Beef	9000	1400	Oltjen *et al.*, 1966
Wheat	Beef	9000	1200	Oltjen *et al.*, 1966
Forage:Corn concentrate				
5:95	Beef	8260	1345	Dinius *et al.*, 1976
34:66	Beef	8235	965	Dinius *et al.*, 1976
64:36	Beef	10,355	770	Dinius *et al.*, 1976
All forage				
Alfalfa hay	Beef	9975	1045	Oltjen *et al.*, 1971; Dinius *et al.*, 1978
Grass hay	Beef	9000	700	Oltjen *et al.*, 1971; Dinius *et al.*, 1978
Alfalfa silage	Beef	9050	755	Bond *et al.*, 1983
Grass silage	Beef	7600	580	Bond *et al.*, 1983

Table 3
Growth Rates of Beef Steers Fed Diets of Different Protein Amounts

Dietary Protein (%)	Feed Intake (kg/Day)	Gain (g/day)
Growing phase		
8	9.46	.98
10	11.07	1.33
12	10.97	1.35
Finishing phase		
8	10.05	.95
10	11.00	1.04
12	10.70	.91

Source: Rumsey, 1985.

trics, the ruminant has the unique ability to digest and utilize the energy in forage and other fibrous and by-product feeds through ruminal fermentation. Because of this ability to utilize feeds that are less expensive and not in direct competition with human food supplies, the ruminant diet contains a large portion of fibrous feedstuffs. Bulk becomes a limitation to intake due to the physical size of the digestive system and rate of passage of undigested material. Thus, for the ruminant, growth rate is often directly related to feed intake and quality or digestibility of the diet.

Dietary protein is usually the second major nutritional contributor to animal growth rates. Most individual feedstuffs do not contain adequate amounts of available protein. Thus, it is necessary to add supplemental protein to a diet. It is usually an expensive part of a given diet and is thus fed to meet animal requirements. Protein density of the diet, up to an optimum level, regulates growth rate. Optimum dietary protein levels for growth and finishing are approximately 22% for broilers (Thomas et al., 1978; National Research Council, 1977), 16% for swine (National Research Council, 1979), 14% for lambs (National Research Council, 1975), and 11% for beef cattle (National Research Council, 1984). Levels of essential amino acids are important in formulating diets for poultry and swine. The relationship between growth rate and dietary protein level for feedlot steers is shown in Table 3.

Age

Under normal conditions of nutritional status and health, growth rate varies in a predictable fashion (Fig. 1). For a short period after birth,

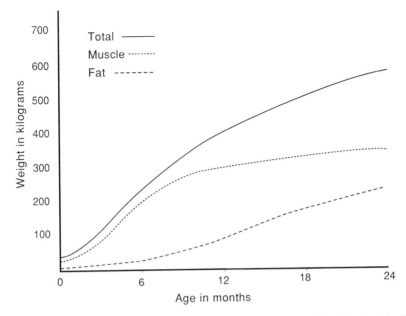

Figure 1 Graphical representation of the change in weight of total animal body, muscle mass, and fat from birth to 24 mo of age.

growth rate is usually slow but gradually increases. This increase becomes exponential over time, reflecting the deposition of protein tissue. As total protein mass reaches the genetic limits of the animal or as fat deposition predominates, growth rate decreases and eventually plateaus at maturity. Thus, as an animal grows, growth rate is a reflection of the type of tissue that is being deposited. Protein tissue or muscle is less energy dense than fat tissue (containing 70% water for muscle and 5% for fat). For growing and finishing animals, the more rapid growth rates are associated with tissue deposition that is predominantly protein, and the slower growth rates are associated with tissue deposition that is predominantly fat.

Growth rate and tissue deposition that we observe with age in animals helps explain variation in growth rate that we see within a species among various breed types. Small-framed, early-maturing breeds have a shorter period in which protein deposition is predominant; that is, fat deposition increases at an earlier age. Large-framed, late-maturing breeds have an extended period during which tissue deposition is predominantly protein in order to achieve greater mature muscle mass. For these animals, fat deposition becomes predominant at a slightly later age. Thus, growth rate over a given time period differs between type of animal because of the genetic and physiological control of the normal

Table 4
Growth Rates of Different Cattle Types

Cattle Type	Growth Rate (g/day)	Reference
Early-maturing		
Angus		
Medium frame	725	Long, 1980
Hereford		
Medium frame	775	Long, 1980
Angus × Hereford		
Medium frame	818	Long, 1980
Hereford × Angus		
Medium frame	785	Long, 1980
Late-maturing		
Charolais	1060	Long, 1980
Charolais	1305	Gregory and Ford, 1983
Gelbveih	1180	Gregory and Ford, 1983
Limousin	1040	Gregory and Ford, 1983

growth curve. An example of the differences one can expect in the growth rate of different types of cattle is shown in Table 4. Within a given breed, genetic variation is exploited in order to change growth rate. Producers who raise breeding stock rate their bulls on growth rate. Another common use of genetic manipulation of growth rate is the use of crossbreeding. Animals with parents of different breeds or genetic lines tended to grow faster (Table 4) (Long, 1980). In these ways, the animal's ability to deposit tissue faster, particularly protein, has been improved.

Growth Promoters

The use of exogenous and endogenous materials that allow the producer to increase growth rate above the normal capabilities of the animal are important tools in current animal production. These materials include primarily antibiotics, ionophores, and hormonal growth promoters. The primary use of antibiotics in animal production is for control of disease processes. In this regard, they improve growth rate because they help maintain a healthier animal. Separate from disease control, some antibiotics appear to promote growth through direct or indirect effects on physiological mechanisms. Current thinking is that antibiotics reduce undesirable microorganisms in the gastrointestinal tract, which increases the amount of available nutrients for absorption and causes a thinning of the gastrointestinal tract wall, which improves absorption of nutrients. Ionophores are materials that were developed originally as coccidiostats. In addition, they have a unique property of changing the end products of ruminal fermentation. A major step in the digestion of feedstuffs by ruminants (cattle and sheep) is anaerobic fermentation of carbon structures to volatile fatty acids (acetic, propionic, and butyric), which are then absorbed as energy sources. Ionophores increase the proportion of propionic acid and decrease the proportion on acetic acid, propionic acid being a more efficient source of energy for the animal than acetic. This unique property of ionophores is being applied to improve growth rate in beef cattle.

Hormonal growth promoters are important tools that are currently used in beef production to increase growth rate in growing and finishing beef cattle. These materials appear to increase the animal's efficiency of depositing protein tissue. That is, during the period of time when protein is the primary tissue being deposited, beef animals treated with these materials deposit approximately 25% more protein mass. A secondary benefit of some of these materials is that feed intake is increased, further increasing growth rate. The percentage of improvement in growth rate of meat-producing farm animals attributable to antibiotics, ionophores, and hormonal growth promoters is shown in Table 5. Overall, these materials account for approximately 18% of our current level of production.

Table 5
Percentage Improved Performance
Attributable to Various Factors

	Attributes (%)			
Animal	Hormonal Growth Promoters	Antibiotics	Ionophores	Total
Calves		25		25
Feedlot cattle	10–15	5	10–12	25
Poultry		5		5
Swine		15		15
Weighted average[a]				18

Sources: Rumsey, 1983; Goodrich *et al.*, 1984.
[a]Based on relative annual production.

Systems

Systems of animal meat production vary among and within species. For both poultry and swine, the systems can be described as intensive feeding programs designed to maximize growth from birth to slaughter. Lamb systems may include: (1) newborn with ewe until weaned (usually with access to a creep feed), followed by a period on pasture, followed by feedlot finishing; (2) the same as (1) but directly from weaning into the feedlot; (3) artificial rearing of newborn lambs followed by feedlot. Cattle systems are more varied: (1) Calves may be marketed at a young age as veal after an intensive period of milk feeding—this is true of a large number of dairy calves; (2) calves may be marketed as baby beef—these are usually beef calves that have been raised intensively on both milk and concentrate feed to about 6 to 8 mo of age; (3) the conventional system for beef cattle is to have calves remain with their mothers (usually on pasture) until weaning at about 6 mo, have them grown on available pasture, stored forage, and/or concentrates to about 12 to 14 mo, and then finish them on high concentrate diets in the feedlot; (4) the same as (3) except move the calves directly to the feedlot after weaning; (5) forage-finishing, where usually the weaned calves are fed primarily pasture or other forages during their growing and finishing period. Growth rates associated with these systems can be estimated from standards published by the National Academy of Science, National Research Council (NRC) for sheep (1975) and for beef (1984) using the information presented relative to feed composition and type and age of animal.

From this chapter it can be concluded that growth is a production parameter that represents a summation of influences on the growing animal, namely, genetic potential, feed intake, age, type of diet, and health as major contributing factors. Growth rate is an important measure by which the producer judges the well-being of his animals, the effectiveness of his system and management, and the planned profitability of his business and is one basis on which to make management decisions.

References

Bereskin, B. (1983). Performance of selected and control lines of Duroc and Yorkshire pigs and their reciprocal crossbred progeny. *J. Anim. Sci.* **57**, 867.
Bond, J., Rumsey, T. S., Berry, B. W., Hammond, A. C., and Dinius, D. A. (1983). Feedlot performance and carcass merit of steers fed silage diets. *J. Anim. Sci.* **57** (Suppl. 1), 279.
Cheeke, P. R. (1983). Utilization of high roughage diets by rabbits. *Maryland Nutrition Conference*, Washington, D.C., p. 32.
Dinius, D. A., Brokken, R. F., Bovard, K. P., and Rumsey, T. S. (1976). Feed intake and

carcass composition of Angus and Santa Gertrudis steers fed diets of varying energy concentration. *J. Anim. Sci.* **42**, 1089.

Dinius, D. A., Goering, J. K., Oltjen, R. R., and Cross, J. R. (1978). Finishing beef steers on forage diets with additives and supplemental lipid. *J. Anim. Sci.* **46**, 761.

Goodrich, R. D., Garrett, J. E., Gast, D. R., Kirick, M. A., Larson, D. A., and Meiske, J. C. (1984). Influence of monensin on the performance of cattle. *J. Anim. Sci.* **58**, 1484.

Gregory, K. E., and Ford, J. J. (1983). Effects of late castration, zeranol and breed group on growth, feed efficiency and carcass characteristics of late maturing bovine males. *J. Anim. Sci.* **56**, 771.

Long, C. R. (1980). Crossbreeding for beef production: Experimental results. *J. Anim. Sci.* **51**, 1197.

Mersmann, H. J., Pond, W. G., and Yen, J. T. (1984). Use of carbohydrate and fat as energy source by obese and lean swine. *J. Anim. Sci.* **58**, 894.

Moran, E. J., Jr., Somers, J., and Larmond, E. (1973). Full-fat soybeans for growing and finishing large white turkeys. 1. Live performance and carcass quality. *Poultry Sci.* **52**, 1936.

National Research Council (1975). Nutrient requirements for domestic animals, no. 5. *In* "Nutrient Requirements for Sheep," 5th rev. ed. National Academy of Sciences–National Research Council, Washington, D.C.

National Research Council (1977). Nutrient requirements for domestic animals, no. 2. *In* "Nutrient Requirements for Poultry," 7th rev. ed. National Academy of Sciences–National Research Council, Washington, D.C.

National Research Council (1979). Nutrient requirements for domestic animals, no. 2. *In* "Nutrient Requirements for Swine," 8th rev. ed. National Academy of Sciences–National Research Council, Washington, D.C.

National Research Council (1984). "Nutrient Requirements of Beef Cattle," 6th rev. ed. National Academy of Sciences–National Research Council, Washington, D.C.

Oltjen, R. R., Putnam, P. A., Williams, E. E., Jr., and Davis, R. E. (1966). Wheat versus corn in all-concentrate cattle diets. *J. Anim. Sci.* **25**, 1000.

Oltjen, R. R., Rumsey, T. S., and Putnam, P. A. (1971). All-forage diets for finishing beef cattle. *J. Anim. Sci.* **32**, 327.

Rand, N. T., Scott, H. M., and Kummerow, F. A. (1958). Dietary fat in the nutrition of the growing chick. *Poultry Sci.* **37**, 1075.

Rumsey, T. S. (1983). Experimental approaches to studying metabolic fate of xenobiotics in food animals. *J. Anim. Sci.* **56**, 222.

Rumsey, T. S. (1985). The role of growth promotants in modern beef production. Maryland Nutrition Conference, Baltimore, p. 50.

Stritzke, D. J., and Whiteman, J. V. (1982). Lamb growth patterns following different seasons of birth. *J. Anim. Sci.* **55**, 1002.

Thomas, O. P., and Bossard, E. H. (1982). Amino acid requirements for broiler males and females. Maryland Nutrition Conference, Washington, D.C., p. 34.

Thomas, O. P., Twining, P. V., Bossard, E. H., and Nicholson, J. L. (1978). Updated amino acid requirements of broilers. Maryland Nutrition Conference, Washington, D.C., p. 107.

Protein Conversion Ratio

Andrew C. Hammond

U.S. Department of Agriculture
Agricultural Research Service
Subtropical Agricultural Research Station
Brooksville, Florida

Introduction

Production of animal protein foods for human consumption has been under considerable scrutiny in recent years for a number of reasons, among which is the question of biological efficiency of protein production (Blaxter, 1973; Breirem et al., 1977; Bywater and Baldwin, 1980; Fitzhugh, 1978; Holmes, 1970; Putnam, 1969; Putnam et al., 1967; Reid, 1970; Winter, 1975). It has been argued that animals compete for plant foods, especially cereal grains and oilseeds, which in today's global society might be more efficiently directed toward human consumption. Many uses of animals other than as food have been identified as being beneficial to humans (Blaxter, 1973; Fitzhugh et al., 1978; McDowell, 1977, 1980), but the fundamental question of biological efficiency remains. Although biological efficiency can be measured in a number of ways (Winter, 1975; Spedding, 1973), a general consensus would be that it is a ratio of output over input in some common unit. In considering the biological efficiency of protein production by farm animals, the output would be some mass unit of protein produced in the form of meat, milk, or eggs and the input would be the same mass unit of protein in the feed consumed. The process involved in production of meat, milk, and eggs can thus be considered a conversion of feed protein to food protein and the efficiency of this process described as a protein conversion ratio.

A general survey of estimated protein conversion ratios among farm animal species or production systems (Table 1) points out that the efficiency of animal protein production can be very low; in fact, the entire range of estimated protein conversion ratios is 4 to 34%. These figures in many ways

Table 1
Protein Conversion Ratios
for Entire Production Systems

Species or Production System	Protein Conversion Ratio[a,b] (%)
Dairy cow	24–34
Laying hen	20–30
Broiler chicken	20–30
Rabbit	11–17
Pig	12–16
Lamb	4–12
Beef steer	6–10

Source: Adapted from Wilson, 1973.
[a]Estimates vary largely due to level of productivity; larger values theoretically can be obtained. [b]Grams protein food produced/grams of feed protein × 100.

are misleading, however. Farm animals, especially ruminants, can utilize forages, by-product feeds, and nonprotein sources of nitrogen, which are all noncompetitive sources of input protein for the production of animal protein foods. Forages, the vegetative parts of plants such as grasses, forbs, legumes, cereals, and browse, supply an average of 60 to 90% of the dietary protein consumed by cattle and sheep in the United States (Putnam *et al.*, 1967) and supply more than 75% of the feed available to ruminants in other regions of the world (Fitzhugh *et al.*, 1978). In addition, vast areas of land not suitable for crop production are grazed by ruminants. Therefore, if forage and other noncompetitive sources of protein are deleted from the inputs required to produce milk and meat from ruminants, then protein conversion ratios of 100% and greater can be obtained (Putnam *et al.*, 1967; Rumsey, 1984). A tabulation of several by-product feeds that contain input sources of feed protein (Table 2) further emphasizes the noncompetitive nature of many farm animal diets. An estimate of the percentage of feed resources for ruminants consisting of crop residues alone is approximately 24% (Fitzhugh *et al.*, 1978). Table 3 compares protein conversions calculated by using total or only competitive protein inputs. Averages are presented for ruminant and nonruminant farm animals, which points out that nonruminants can be more efficient converters of total feed protein inputs but that a larger potential for converting noncompetitive feed protein inputs goes to ruminant farm animals.

Tables 4 and 5 show the relationship between varying roughage levels and the conversion of feed protein to protein gain in beef steers and heifers, respectively. Protein conversions are higher in Tables 4 and 5 compared to the general trends for beef production in Tables 1 and 3 because output is total protein gain rather than protein only in edible meat. Furthermore, Tables 4 and 5 consider only one phase (growing/finishing) of the entire beef production system. Nevertheless, the relative effects of type of diet on protein conversion are well illustrated. Also shown in these tables is the effect of reducing total feed protein inputs by replacing one-third of the required feed protein with nonprotein sources of nitrogen. This level of nonprotein use in cattle diets is consistent with most recommendations, but under experimental conditions it has been shown that nonprotein sources of nitrogen can entirely replace the natural protein in the diet of ruminants (Oltjen, 1969; Virtanen, 1966). Some examples of nonprotein sources of feedstuff nitrogen are listed in Table 6.

It should be concluded from the data presented in this chapter that the contribution of farm animal production to the supply of protein foods for human consumption should not be discounted due to relative biological efficiency of protein production. Consideration should be given to the noncompetitive nature of many of the feed inputs that go into animal production and the ability of ruminants to convert nonprotein sources of nitrogen into edible protein.

Table 2
By-Products and Crop Residues Containing Sources of Protein[a]

Feedstuff	Crude Protein Content (%)
Avocado seed meal	20
Beans, cull navy	24
Brewer's grains, wet	26
Brewer's grains, dried	28
Brewer's yeast, dried	48
Cattle manure, dried	15
Corn gluten feed	27
Corn gluten meal	43
Cottonseed meal	45–46
Distillers' grains	30
Distillers' solubles, dried	30
Feather meal, hydrolyzed	91
Garbage, cooked municipal	16
Grain screenings	15
Hop vine silage	15
Hops, spent	22
Blood meal	87
Meat and bone meal	54
Linseed meal	40
Peanut meal	52
Peas, cull	25
Poultry litter, dried	30
Poultry manure, dried	30
Rapeseed meal	41
Safflower meal	22
Safflower meal, dehulled	49
Soybean meal	52–56
Sunflower meal	50
Sunflower meal with hulls	32
Tomato pomace, dried	23
Turnip tops (purple)	16
Wheat bran	18
Wheat middlings	18
Wheat mill run	17
Wheat shorts	20
Whey, dried	16

Source: Partially adapted from Preston, 1979.
[a]Examples listed are of noncompetitive feedstuffs.

Table 3
Comparisons of Protein Conversion Ratios

Production Unit	Protein Conversion Ratio Using Total Protein Input	Protein Conversion Ratio Using Competitive Protein Input
Ruminant		
Beef	14.1	43.5
Sheep	6.9	52.6
Dairy	24.4	105.3
Average	11.7	49.9
Nonruminant		
Swine	16.9	18.2
Broilers	25.6	40.0
Layers	25.6	45.5
Average	21.8	29.4

Source: Adapted from Rumsey, 1984; original data from Cunha, 1982; Wilson, 1980.

Table 4
Estimated Effects of Percentage of Dietary Roughage and Animal Weight on Protein Conversion Ratios in Growing Steers

Roughage in diet (%)	Animal weight (kg)	Daily gain (kg)	Composition of gain (% protein)	Total protein requirement (kg)	Total protein conversion ratio[a] (%)	Whole protein conversion ratio[b] (%)
70–80	100	0.5	18.0	0.36	25.0	37.3
50–60	100	0.7	18.0	0.40	31.5	47.0
25–30	100	0.9	18.0	0.46	35.2	52.6
15	100	1.1	18.0	0.49	40.4	60.3
55–65	300	0.9	13.5	0.81	15.0	22.4
20–25	300	1.1	13.5	0.82	18.1	27.0
15	300	1.3	13.5	0.83	21.1	31.6
45–55	500	0.9	9.0	0.95	8.5	12.7
20–25	500	1.1	9.0	0.96	10.3	15.4
15	500	1.2	9.0	0.96	11.3	16.8

Source: Derived from National Research Council, 1976.
[a]Grams of protein gain/grams of feed protein regardless of source of nitrogen × 100. [b]Calculated as in [a] but assuming feed protein input reduced one-third by the use of nonprotein nitrogen.

Table 5
Estimated Effects of Percentage of Dietary Roughage and
Animal Weight on Protein Conversion Ratios in Growing Heifers

Roughage in Diet (%)	Animal Weight (kg)	Daily Gain (kg)	Composition of Gain (% Protein)	Total Protein Requirement (kg)	Total Protein Conversion Ratio[a] (%)	Whole Protein Conversion Ratio[b] (%)
70–80	100	0.5	18.0	0.37	24.3	36.3
50–60	100	0.7	18.0	0.42	30.0	44.8
25–30	100	0.9	18.0	0.48	33.8	50.4
< 15	100	1.1	18.0	0.53	37.4	55.8
100	300	0.3	12.5	0.63	5.6	8.4
80–90	300	0.5	12.5	0.67	9.3	13.9
55–65	300	0.7	12.5	0.67	13.1	19.5
35–45	300	0.9	12.5	0.70	16.1	24.0
20–25	300	1.1	12.5	0.78	17.6	26.3
< 15	300	1.2	12.5	0.79	19.0	28.3

Source: Derived from National Research Council, 1976.
[a]Grams of protein gain/grams of feed protein regardless of source of nitrogen × 100. [b]Calculated as in [a] but assuming feed protein input reduced one-third by the use of nonprotein nitrogen.

Table 6
Examples of Nonprotein Sources
of Nitrogen Used as Feedstuffs
for Ruminants

Feedstuff[a]	Crude Protein Content[b]
Biuret[c]	219
Diammonium phosphate[d]	115
Monoammonium phosphate[d]	74
Urea[d]	287

[a]Feed grade products may not be chemically pure, so crude protein content may be less than calculated for pure chemical; see [b]. [b]Average nitrogen concentration in feed protein assumed to be 16%; crude protein content calculated by multiplying nitrogen content by 6.25. [c]From Tiwari et al., 1973. [d]From Preston, 1979.

References

Blaxter, K. L. (1973). Increasing output of animal production. *In* "Man, Food, and Nutrition" (M. Rechcigl, Jr., ed.), p. 127. CRC Press, Cleveland Ohio.

Breirem, K., Homb, T., and Vik-Mo, L. (1977). Production of animal protein in view of human protein demand. *In* "Proceedings of the Second International Symposium on Protein Metabolism and Nutrition," EAAP 22, p. 154. Centre for Agricultural Publishing and Documentation, Wageningen.

Bywater, A. C., and Baldwin, R. L. (1980). Alternative strategies in food-animal production. *In* "Animals, Feed, Food and People," AAAS 42, (R. L. Baldwin, ed.), Chap. 1. Westview Press, Inc., Boulder, Colorado.

Cunha, T. J. (1982). The animal as a food source for man. *Feedstuffs* **54**, 18.

Fitzhugh, H. A. (1978). Bioeconomic analysis of ruminant production systems. *J. Anim. Sci.* **46**, 797.

Fitzhugh, H. A., Hodgson, H. J., Scoville, O. J., Nguyen, T. D., and Byerly, T. C. (1978). "The Role of Ruminants in Support of Man." Winrock International, Morrilton, Arkansas.

Holmes, W. (1970). Animals for food. *Proc. Nutr. Soc.* **29**, 237.

McDowell, R. E. (1977). "Ruminant Products: More than Meat and Milk." Winrock International, Morrilton, Arkansas.

McDowell, R. E. (1980). The role of animals in developing countries. *In* "Animals, Feed, Food and People," AAAS 42, (R. L. Baldwin, ed.), Chap. 6. Westview Press, Inc., Boulder, Colorado.

National Research Council (1976). "No. 4, Nutrient Requirements of Beef Cattle," 5th rev. ed. National Academy of Sciences, Washington, D.C.

Oltjen, R. R. (1969). Effects of feeding ruminants non-protein nitrogen as the only nitrogen source. *J. Anim. Sci.* **28**, 673.

Preston, R. L. (1979). Typical composition of feeds for cattle and sheep, 1979–1980. *Feedstuffs* **51**, 3A.

Putnam, P. A. (1969). What are the prospects for milk products as sources of protein? *J. Dairy Sci.* **52**, 419.

Putnam, P. A., Moore, L. A., and Bayley, N. D. (1967). A challenge to ruminant agriculture. in *Proc. Cornell Nutr. Conf.*, p. 44. Cornell University, Ithaca, New York.

Reid, J. T. (1970). Will meat, milk, and egg production be possible in the future? in *Proc. Cornell Nutr. Conf.*, p. 50. Cornell University, Ithaca, New York.

Rumsey, T. S. (1984). Monensin in cattle: Introduction. *J. Anim. Sci.* **58**, 1461.

Spedding, C. R. W. (1973). The meaning of biological efficiency. *In* "The Biological Efficiency of Protein Production" (J. G. W. Jones, ed.), Chap. 3. Cambridge University Press, Cambridge.

Tiwari, A. D., Owens, F. N., and Garrigan, U. S. (1973). Metabolism of biuret by ruminants: *In vivo* and *in vitro* studies and the role of protozoa in bimetolysis. *J. Anim. Sci.* **37**, 1396.

Virtanen, A. J. (1966). Milk production of cows on protein-free feed. *Science* **153**, 1603.

Wilson, P. N. (1980). The availability of feeds for animal production systems. *In* "IV World Conference on Animal Production" (L. S. Verde and A. Fernandez, eds.), p. 133. Asociation Argentina de Produccion Animal, Buenos Aires.

Wilson, P. N. (1973). The biological efficiency of protein production by animal production enterprises. *In* "The Biological Efficiency of Protein Production" (J. G. W. Jones, ed.), Chap. 13. Cambridge University Press, Cambridge.

Winter, G. R. (1975). "Protein Efficiency in Canada." Canadian Livestock Feed Board, Montreal, Canada.

Production Systems

L. W. Smith

U.S. Department of Agriculture
Agricultural Research Service
National Program Staff
Beltsville, Maryland

Introduction

The production of farm animals for food and fiber—that is, for meat, milk, and wool—is dependent upon adequate resources, particularly the availability of land, water, and energy. Animal production systems have evolved to accommodate climatic variation. Animal agriculture is compatible with human environmental quality objectives. Some production systems achieve and maintain a more desirable environmental quality than alternative economic uses of land. Examples include managed grazing of rangeland and pastureland and production of forages on land too steep for row-crop production. Modern animal production systems usually comprise several management practices in various combinations. For instance, dairy farmers have several alternatives in feeding and management of the dairy herd; extent of using farm-grown forages and grains; options in acquiring herd replacements; methods for winter storage of feeds; types of shelter and housing; and methods of management and use of animal manure. The number of livestock (beef cattle, dairy cattle, and sheep) is shown by state in Table 1. Livestock are adaptable to widely varying environmental conditions as indicated by their presence in each state of the United States.

Resources

Land

The amount of land needed to support livestock production varies with the class of livestock, production system (or management objective), climate, soil conditions and topography, and design of physical facilities. From a fraction of an acre on diversified farms to more than 100 acres of land per animal are needed on some rangelands. In the more intensive systems, livestock are fed feedstuffs transported to them from cultivated lands. In the highly intensive production systems, the only land directly

Table 1
Number of Livestock by State

	Beef Cows[a] 1982	Calves Born[b] 1981	Cattle and Calves on Feed January 1, 1982 (Thousands)	Milk Cows 1982	Stock Sheep 1982	Sheep on Feed 1982
Northeastern						
Connecticut	7	50	—	49	6.3	—
Delaware	5	12	—	10	—	—
Maine	16	56	—	59	15	—
Maryland	92	180	16	122	19	—
Massachusetts	10	47	—	47	8	—
New Hampshire	6	32	—	30	7.7	—
New Jersey	14	45	1	41	9.7	—
New York	129	960	12	921	70	—
Pennsylvania	247	850	75	730	125	—
Rhode Island	.6	4.0	—	3.8	—	—
Vermont	12	180	—	190	11	—
Total	538.6	2,416	104	2,202.8	271.7	—
Southern						
Alabama	950	870	34	60	—	—
Arkansas	1,016	980	11	84	—	—
Florida	1,220	1,150	90	190	—	—
Georgia	901	850	40	131	—	—
Kentucky	1,050	1,210	30	240	25	—
Louisiana	758	600	11	102	9.9	—
Mississippi	953	880	10	97	—	—
North Carolina	436	480	28	134	7.7	—
Oklahoma	2,340	2,160	270	110	90	15
South Carolina	309	280	20	48	—	—
Tennessee	1,010	1,080	20	215	10	—
Texas	5,925	5,400	1,660	325	2,200	200
Virginia	668	750	51	172	170	—
West Virginia	279	260	4	36	110	—
Total	17,815	16,950	2,279	1,944	2,622.6	215
North Central						
Illinois	808	890	480	234	175	20
Indiana	463	600	240	207	129	9
Iowa	1,735	1,950	1,130	389	400	85
Kansas	1,826	1,770	1,110	124	140	60
Michigan	214	500	135	396	106	24
Minnesota	585	1,370	350	905	275	60
Missouri	2,205	2,370	90	245	115	18
Nebraska	2,148	1,970	1,640	122	140	85
North Dakota	978	1,050	36	92	230	50
Ohio	388	110	125	382	260	53
South Dakota	1,595	1,720	335	160	700	50
Wisconsin	257	2,020	112	1,830	110	15
Total	13,202	16,920	5,783	5,086	2,780	529

(*continued*)

Table 1   (*Continued*)

	Beef Cows[a] 1982	Calves Born[b] 1981	Cattle and Calves on Feed January 1, 1982 (Thousands)	Milk Cows 1982	Stock Sheep 1982	Sheep on Feed 1982
Western						
Alaska	3	3.2	—	1.1	3.9	—
Arizona	275	280	330	80	320	57
California	1,160	1,710	581	940	1,010	200
Colorado	945	900	750	75	480	230
Hawaii	80	72	14	13	—	—
Idaho	562	750	232	168	470	28
Montana	1,604	1,580	53	29	600	16
Nevada	359	290	24	16	116	13
New Mexico	565	600	127	55	595	20
Oregon	730	750	78	97	440	100
Utah	364	375	48	86	610	26
Washington	443	530	160	207	83	—
Wyoming	673	650	52	12	1,000	130
Other states	—	—	4	—	—	—
Total	7,763	8,490.2	2,453	1,779.1	5,727.9	820
U.S. total	39,319	44,776	10,619	11,012	11,402	1,564

Source: U.S. Dept. of Agriculture, 1983.
[a]Cows and heifers that have calved. [b]Includes calves born to beef cows and dairy cows.

used by animals is land occupied by the feedlot or confinement housing facility. Usually, intensive systems use high-energy density feeds (grain), protein supplements, and vitamin and mineral supplements with a minimum of forage. Intensive systems require more sophisticated management to avoid adverse impact on the environment. There is a complex interaction among and between cropland, improved pasture, and rangeland for some animal production systems. Use of land in the United States is shown for 1977 in Fig. 1. Rangeland and pastureland exceeded cropland by more than 100 million acres in 1977. If federal rangeland were included, this would be even more striking.

Water

The quantity and quality of water is important for all animal production systems. The direct consumption of water needed to satisfy the thirst of the animal is small compared to large quantities of water needed to grow crops for feed. For 1977, 58 million acres were irrigated (Fig. 2). Drinking water for livestock must be of good quality and free of toxic materials

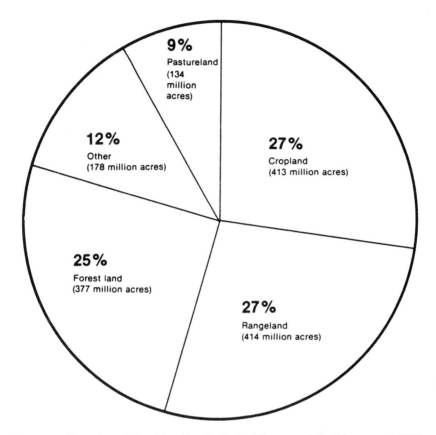

Figure 1 Use of nonfederal land in the United States and Caribbean area, 1977.

and pollutants. However, livestock can be raised on water of lower quality than is needed for humans.

Energy

The activities associated with production of food and fiber on the farm are all involved in using or managing different forms of energy. In the broadest sense, all forms of crop production capture solar energy. Animals utilize plant materials and convert them into products more acceptable or appealing to the consumer.

Animal agriculture requires energy expenditures in the form of human labor; raw materials for chemicals; fertilizer, pesticides and herbicides; fuel for tilling and managing land with machines; storage structures for crops used for feed; livestock shelters; irrigation water in

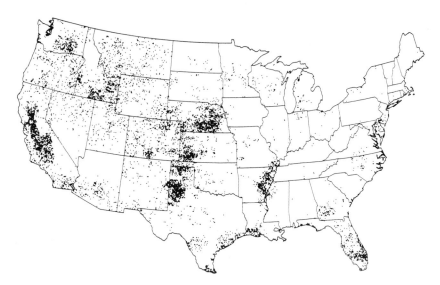

Figure 2 Irrigated acreage in the United States. One dot equals 8000 acres where irrigation facilities are in place. Total irrigated land equals 58 million acres, 1977.

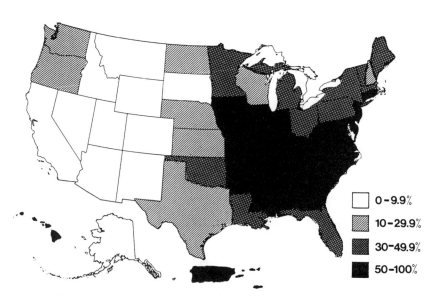

0-9.9%
10-29.9%
30-49.9%
50-100%

Figure 3 Percentage of nonfederal forage-producing land managed as pastureland, by state.

rainfall-deficient areas; food processing; and transport of products to market. The maintenance energy of animals in some production systems is large relative to energy used for productive purposes (i.e., growth, gestation, milk production) and must be from low-cost feed such as forage grazed from range, pasture, or crop residue to provide margins for profit. Fig. 3 shows the percentage of forage-producing land managed as pastureland by state. Production systems utilizing pasture are more culturally energy efficient, as less fossil fuel energy is used per unit of production. Animal agriculture is adaptable to very low energy inputs, but output will be low.

About 3% of total U.S. energy usage is by farming, and its cost is significant in the total cost of production (CAST, 1977). Energy associated with production of fertilizer and other agricultural chemicals uses the most fossil fuel energy, followed by fuel for operating machinery. Several times more energy is used in getting animal products processed and distributed than is used for on-farm production.

Ruminant animals have the unique capability to digest forages, utilize nonprotein nitrogen, and convert them into human food. Ruminants include beef cattle, dairy cattle, sheep, and goats, as well as other species such as deer and bison. Their digestive tract is anatomically adapted to microbiologically ferment plant materials and change them into usable nutrients. This feature allows them to convert cellulosic plant materials and nitrogen to milk and meat that would otherwise be unavailable for human consumption. In order to achieve higher performance and to improve the quality of meat, some cereal grains are often needed to supplement forages.

The proportion of dietary energy provided by grains, high-protein feeds, and other grain by-products is shown in Table 2 for dairy cattle,

Table 2
Dietary Energy Provided by Certain Feedstuffs
to Various Livestock, 1972–73

Feedstuffs	Percentage of derived dietary energy					
	Dairy cattle	Beef cattle	Pigs	Poultry	Other livestock	All livestock
Grains[a]	25.2	21.8	72.0	62.5	47.0	35.6
High-protein feeds[b]	5.3	2.6	11.6	29.5	9.0	7.5
Other by-products[c]	4.7	1.2	2.1	5.4	3.8	2.5
Roughage[d]	64.8	74.4	14.3	2.6	40.2	54.4

[a]Cereal and sorghum grains. [b]Includes oilseed meals, slaughter by-products, and grain-protein by-products. [c]Includes gain-milling by-products. [d]Includes pasture, range, hay, silage, and straw.

Table 3
Metabolizable Energy and Grains Fed to Livestock
and Their Contributions to Human Food Energy, 1977

	Energy Used by Livestock (%)	Grains Used by Livestock (%)	Human Food Energy from Livestock (%)
Poultry	7	27	9
Swine	10	32	30
Dairy	24	18	40
Beef	32	17	18
Sheep and goats	12	2	3
Other[a]	15	4	(negligible)

Source: Wheeler *et al.*, 1981.
[a]Primarily draft animals.

beef cattle, pigs, and poultry. Only 21.8% of the total dietary energy is from grains for beef cattle. Roughages are not effectively utilized by monogastric farm animals. Sheep and goats, if properly managed, are more efficient than larger ruminants for meat production because they attain market weight in a shorter period of time and have a higher frequency of twinning than cattle. Animals that produce an acceptable product directly from range or pasture with minimal input of grain feeding have an advantage in that they do not compete for grain or acreage that could produce food directly for people. A disadvantage is that rangeland and pastureland must be managed well in order to prevent damage.

Another perspective is provided in Table 3 on the relative competitiveness of livestock for total feed energy, grain, and their use for human food energy. Dairy cattle use 24% of all metabolizable energy used by livestock, consume 18% of the grain, and provide 40% of the energy humans obtain from livestock products. Ruminants (cattle, sheep, and goats) provide 61% of all livestock of livestock-furnished human energy but use only 37% of the grain consumed by livestock on a world basis (Wheeler *et al.*, 1981).

Factors Affecting a Production System

In Table 4 production systems are broadly categorized as intensive and extensive. In general, intensive systems are those in which animals are

Table 4
Categorization of Some Animal Production Systems

System	Type	Size (Head)	Production Objective	Rate of Production
Intensive				
Beef	Farmer-feeder	< 1000	Finished cattle	1.0 ± .20 kg/day
	Commercial-feedlot	> 1000	Finished cattle	1.0 ± .20 kg/day
Sheep	Commercial-feedlot	> 1,000	Market lamb	.40 ± .10 kg/day
	Farm-flock	< 1,000	Market lamb and wool	.40 ± .10 kg/day
Dairy cattle	Dry lot	> 1,000	Milk[a]	> 10,000 kg/year
	Free-stall barn	> 50	Milk	> 10,000 kg/year
	Stanchion barn	50 to 200	Milk	> 10,000 kg/year
Extensive				
Beef cow-calf	Range	> 1,000	Feeder cattle herd replacements	One offspring per breeding female annually
	Pasture	100	Feeder cattle herd replacements	One offspring per breeding female annually
Sheep	Range	> 1,000	Feeder lambs	One and one-half lambs per breeding female annually

[a]Calves and herd replacements represent secondary and tertiary production objectives.

kept in confined housing and fed diets to achieve accelerated production. Therefore, extensive systems are those maximizing use of grazed forage—crop residue with only limited supplementation of essential nutrients. Table 5 summarizes some of the factors affecting choice and use of a production system.

Although some kinds of animal production systems are more prevalent in some parts of the country, a generalization for locating their distribution is not possible. The reader is referred to more detailed descriptions in textbooks on production of beef cattle (Newman, 1977), dairy cattle (Foley et al., 1973), and sheep (Scott, 1970).

In some intensive production systems, a facility is often needed as part of an animal production system to provide shelter or to control and manage animals. Table 6 lists some kinds of facilities used in the production of livestock, including options for handling manure as a solid, slurry, or liquid. Manure handling and management are expenses charged

Table 5
Factors Influencing Choice of Animal Production Systems

Environmental	Animal	Economic	Operator	Government
Climate Rainfall—total and seasonal distribution Topography Humidity Wind (seasonal distribution) Latitude Maximum–minimum temperature Land Value for alternative uses	Species Breed Production objective (pure-bred breeding stock, milk, meat, etc.) Shelter–housing	Profit Alternative—market crops by feeding to animals instead of selling grain and forage Cost of energy Distance from market	Personal preference Cultural background Profit incentive	Environmental constraints—regulation against mismanagement of wastes, odor, and water pollution

Table 6
Alternative Livestock Management Systems

Livestock Commodity	Alternative Physical Plants	Manure-Handling Alternatives				
		Solid		Slurry		Liquid
		With Bedding	Without Bedding	With Bedding	Without Bedding	Without Bedding
Dairy	Stanchion Loose housing Free stall	Stack Bunker Pack	Paved lot Mounds Debris basin Pasture	Earthen pit	Pit Tank	Holding pond Lagoon
Beef	Unpaved lot Paved lot Confined housed	Pack Bunker	Mound Debris basin Paved lot Unpaved lot		Earthen pit	Lagoon Oxidation ditch Holding pond
Swine	Farrowing Confined housed Growing finishing				Pit Pit	Lagoon Lagoon
	Unpaved lot Paved lot Confined housed Pasture	Pack	Mounds Paved lot Debris basin		Pit Tank	Lagoon Oxidation ditch Holding pond Earthen pit
Sheep	Unpaved lot Pasture Confined housed	Pack Stack	Mound		Pit	Lagoon

Source: Gilbertson et al., 1979.

Table 7
Estimated Quantities and Constituents of Livestock Manure[a]

Livestock	Manure Quantity (Metric Tons/Animal Year)		Volatile Solids (% Dry Weight)	Total Solids (% Wet Basis)	Constituents (kg/Animal yr)									
	Wet	Dry			N	P	K	Fe	Zn	Mn	Cu	Ca	Na	Mg
Dairy	13.5	1.7	81.0	12.7	55.6	9.3	44.5	0.7	0.10	0.10	0.04	32.5	6.5	9.9
Beef	6.1	0.7	87.0	11.6	27.8	8.1	17.6	0.9	0.09	0.09	0.00	5.2	1.9	2.6
Swine	2.1	0.2	81.5	9.2	14.5	3.3	5.2	0.1	0.90	0.30	0.04	5.1	0.8	1.3
Sheep	0.6	0.2	85.0	25.0	7.0	1.6	4.9	0.2	0.02	0.02	0.00	0.4	0.3	0.3

Source: Gilbertson et al., 1979.

[a] Average animal weight was as follows: dairy and beef, 453 kg; swine, 91 kg; sheep, 45 kg. Fe, Zn, Mn and Cu for sheep manure were calculated by proportion using beef cattle manure values.

against animal production. Manure production and its nutrient content are given in Table 7. Effective management and use of manure in crop production is usually the most suitable method for using this resource. Alternative uses of manure, including recycling for feed and use as substrate for methane production, may be more economical than its use for crop production (Smith and Wheeler, 1979).

References

CAST (1977). Energy conservation in agriculture. Council for Agricultural Science and Technology, Special Pub. No. 5.

Foley, R. C., Bath, D. L., Dickinson, F. N., and Tucker, H. A. (1973). "Dairy Cattle: Principles, Practices, Problems, Profits." Lea and Febiger, Philadelphia, Pennsylvania.

Gilbertson, C. B., Van Dyne, D. L., Clanton, C. J., and White, R. K. (1979). Estimating quantity and constituents in livestock and poultry manure residue as reflected by management systems. *Transactions of ASAE* **22, (No. 3)**, 602.

Newman, A. L. (1977). "Beef Cattle," 7th ed. John Wiley and Sons, New York.

Scott, G. E. (1970). "The Sheepman's Production Handbook," 1st ed. SID Program, Denver, Colorado.

Smith, L. W., and Wheeler, W. E. (1979). Nutritional and economic value of animal excreta. *J. Anim. Sci.* **48,** 144.

U.S. Department of Agriculture (1981). Soil and Water Resources Conservation Act. Soil, water and related resources in the United States: Status, conditions, and trends, 1980, *Appraisal,* Part 1.

U.S. Department of Agriculture (1983). "Agricultural Statistics." Government Printing Office, Washington, D.C.

Wheeler, R. O., Cramer, G. L., Young, K. B., and Ospina, E. (1981). "Technical Report: The World Livestock Product, Feedstuff and Feed Grain System." Winrock International, Moulton, Arkansas.

Product/
Utilization

By-Products

Anthony W. Kotula

Meat Science Research Laboratory
Agricultural Research Service
U.S. Department of Agriculture
Beltsville, Maryland

Introduction

Slaughter animals, as described in the other chapters of this book, are raised for the ultimate sale of meat and meat products. Just as the producer seeks to implement economies in production practices, the packer is interested in finding markets for each edible and inedible part of the live animal he has purchased. It is only through diligent and innovative utilization of the complete animal that the industry can expect to maintain a firm marketing position and compete favorably with animal products from countries with lower production costs or with nonanimal proteins, as in soybeans or grains. The second consideration facing the packer is that any by-products which are not marketable become an added expense because of loss of potential revenue and disposal costs. The third consideration is the challenge facing the packer to be an effective steward of the natural resources placed at his disposal. Therefore, in this section, we shall discuss the intrinsic value of slaughter animals as they arrive at the packer for conversion into meat and edible and inedible by-products.

Tables 1, 2, and 3 show the by-products and their use from cattle (Swift Agriculture Bulletin 6), hogs (Swift Agriculture Bulletin 10), and lambs (Swift Agriculture Bulletin 19). Additional extensive information on animal by-products can be found in books by Levie (1963), Romans and Ziegler (1974), and Forrest *et al.* (1975), the proceedings of the session on by-products utilization at the 29th Reciprocal Meat Conference of the American Meat Science Association, and an article by Dailey (1977) on the use of pharmaceuticals and medicinals of animal origin. Rather than reiterate what has been published in those excellent references, the remainder of this section will focus on the principle by-products whose use is changing or those that afford the greatest opportunities or challenges for new product development and use. These include manure from pens at the meat establishment, blood, hides and skins, and bones. Dressing percentages and the amount of edible meat from the various slaughter animals also will be discussed. Though the conversion of inedible packinghouse raw materials into edible protein materials for human consumption holds great potential, the procedures have not yet been researched adequately to make this concept economically feasible. Articles by Olson (1970), Levin (1970), and Areas and Lawrie (1984) describe the concept of recovering high-quality protein concentrate from packinghouse raw materials.

Manure

During transport and while in holding pens, livestock continue to produce manure that is of economic importance as fertilizer, feed supplements, or energy sources. Historically, animal manure has been collected and used to supply nitrogen, potassium, phosphorus, and calcium to the soil for plant production, as shown in Table 4 (Smith and Wheeler, 1979). Some farmers are paid to haul away manure, whereas in some instances, farmers pay a nominal fee to have it delivered (*Feedstuffs*, 1985). On the average, a farmer can spread about 10 tons of manure per acre in one season. As a feed supplement, manure is worth about four times more than its value as a fertilizer. Table 5 (Smith and Wheeler, 1979) provides values for the nutrient composition of manure from poultry, cattle, and swine. Table 6 (Smith, 1977), which provides an estimate of dry matter produced by selected classes of slaughter animals, may be used to estimate the amounts of manure available for use or disposal from holding pens or slaughter establishments. The amino acid content of manure as presented in Table 7 (Smith, 1973) is provided to aid in formulating rations, when manure is to be used in that manner. The incorporation of manure into feed rations must be carried out in stages over a 30-day period, as explained by Copeland (1982). He recommended providing calves with 50% litter, 50% corn, and grazing ad libidum. After the cattle reach about 700 lb, the litter should be reduced to about 40% of the ration.

The potential concern about the presence of chemical residues in

Table 1
Cattle By-Products

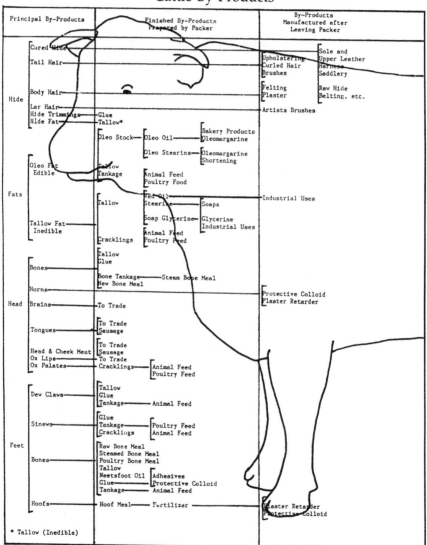

Principal By-Products	Finished By-Products Prepared by Packer	By-Products Manufactured after Leaving Packer
Hide — Cured Hide, Tail Hair, Body Hair, Lar Hair, Hide Trimmings, Hide Fat	Glue, Tallow*	Upholstering, Curled Hair, Brushes, Felting, Plaster, Artists Brushes — Sole and Upper Leather, Harness, Saddlery, Raw Hide Belting, etc.
Fats — Oleo Fat Edible; Tallow Fat Inedible	Oleo Stock — Oleo Oil — Bakery Products, Oleomargarine; Oleo Stearins — Oleomargarine, Shortening; Tallow Tankage — Animal Feed, Poultry Food; Tallow Stearine — Red Oil, Soaps; Soap Glycerine — Glycerine Industrial Uses; Cracklings — Animal Feed, Poultry Feed; Tallow, Glue	Industrial Uses
Head — Bones, Horns, Brains, Tongues, Head & Cheek Meat, Ox Lips, Ox Palates	Bone Tankage — Steam Bone Meal; New Bone Meal; To Trade; To Trade, Sausage; To Trade, Sausage; To Trade; Cracklings — Animal Feed, Poultry Feed	Protective Colloid, Plaster Retarder
Feet — Dew Claws, Sinews, Bones, Hoofs	Tallow, Glue, Tankage — Animal Feed; Glue, Tankage — Poultry Feed, Animal Feed, Cracklings; Raw Bone Meal, Steamed Bone Meal, Poultry Bone Meal, Tallow, Neetsfoot Oil, Glue — Adhesives, Protective Colloid, Tankage — Animal Feed; Hoof Meal — Fertilizer	Plaster Retarder, Protective Colloid

* Tallow (Inedible)

Source: Swift and company, c. 1950

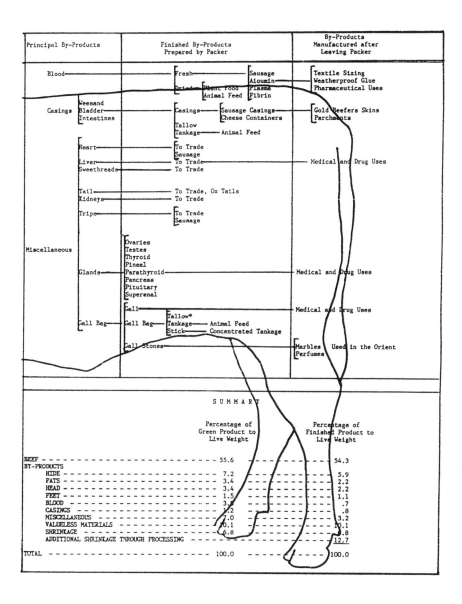

Principal By-Products	Finished By-Products Prepared by Packer	By-Products Manufactured after Leaving Packer

Blood ——— Fresh ——— Sausage / Aloumin → Textile Sizing / Weatherproof Glue / Pharmaceutical Uses

Dried — Plant Food / Animal Feed — Plasma / Fibrin

Casings — Weesand / Bladder / Intestines — Casings — Sausage Casings / Cheese Containers → Gold Beefers Skins / Parchments

Tallow / Tankage — Animal Feed

Miscellaneous:

Heart ——— To Trade / Sausage

Liver ——— To Trade → Medical and Drug Uses

Sweetbreads ——— To Trade

Tail ——— To Trade, Ox Tails

Kidneys ——— To Trade

Tripe ——— To Trade / Sausage

Glands — Ovaries / Testes / Thyroid / Pineal / Parathyroid / Pancreas / Pituitary / Superenal → Medical and Drug Uses

Gall Bag — Gall — Medical and Drug Uses

Gall Bag — Tallow / Tankage — Animal Feed / Stick — Concentrated Tankage

Gall Stones → Marbles Used in the Orient / Perfumes

SUMMARY

	Percentage of Green Product to Live Weight	Percentage of Finished Product to Live Weight
BEEF	55.6	54.3
BY-PRODUCTS		
HIDE	7.2	5.9
FATS	3.4	2.2
HEAD	3.4	2.2
FEET	1.5	1.1
BLOOD	3.?	.7
CASINGS	?.2	.8
MISCELLANEOUS	?.0	3.2
VALUELESS MATERIALS	?0.1	?0.1
SHRINKAGE	?6.8	?.8
ADDITIONAL SHRINKAGE THROUGH PROCESSING		12.7
TOTAL	100.0	100.0

Table 2
Hog By-Products

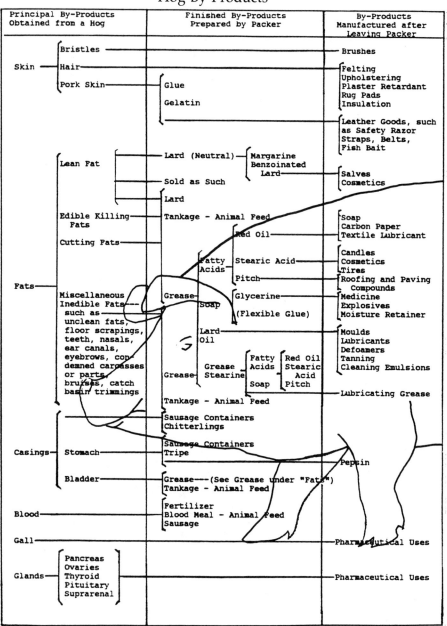

Principal By-Products Obtained from a Hog	Finished By-Products Prepared by Packer	By-Products Manufactured after Leaving Packer
Skin — Bristles		Brushes
Hair		Felting, Upholstering, Plaster Retardant, Rug Pads, Insulation
Pork Skin	Glue, Gelatin	Leather Goods, such as Safety Razor Straps, Belts, Fish Bait
Fats — Lean Fat	Lard (Neutral) — Margarine, Benzoinated Lard	Salves, Cosmetics
	Sold as Such	
	Lard	
Edible Killing Fats	Tankage – Animal Feed	
Cutting Fats	Red Oil	Soap, Carbon Paper, Textile Lubricant
	Fatty Acids — Stearic Acid	Candles, Cosmetics, Tires
	Pitch	Roofing and Paving Compounds
	Grease, Soap — Glycerine	Medicine, Explosives, Moisture Retainer
	(Flexible Glue)	
Miscellaneous Inedible Fats such as unclean fats, floor scrapings, teeth, nasals, ear canals, eyebrows, condemned carcasses or parts, bruises, catch basin trimmings	Lard Oil	Moulds, Lubricants, Defoamers, Tanning, Cleaning Emulsions
	Grease Stearine — Fatty Acids [Red Oil, Stearic Acid, Pitch], Soap	Lubricating Grease
	Tankage – Animal Feed	
Casings — Stomach	Sausage Containers, Chitterlings	
	Sausage Containers, Tripe	Pepsin
Bladder	Grease––(See Grease under "Fats"), Tankage – Animal Feed	
Blood	Fertilizer, Blood Meal – Animal Feed, Sausage	
Gall		Pharmaceutical Uses
Glands — Pancreas, Ovaries, Thyroid, Pituitary, Suprarenal		Pharmaceutical Uses

Source: Swift and company, c. 1955

Principal By-Products Obtained from a Hog	Finished By-Products Prepared by Packer	By-Products Manufactured after Packer
Head — Tongue	Sausage / Potted Tongue / Canned Pickled Tongue	
Ears	Sausage / Sold as Such	
Lips	Sausage / Sold as Such	
Snout	Sausage / Sold as Such	
Cheek & Head Meat	Sausage / Sold as Such	
Brain	Canned	
Bones	Lard / Glue / Steam Bone Fertilizer / Tankage — Animal Feeds / Grease	(See Grease under "Fats")
Mis'l — Feet	Lard / Sold as Such / Pickled and Dry Salt / Grease / Glue / Tankage — Animal Feed	(See Grease under "Fats")
Tail	Sold as Such	
Heart	Sold as Such / Sausage	
Liver	Sold as Such / Sausage	Liver Extract Pharmaceutics
Lungs	Tankage — Animal Feed	
Kidneys	Sold as Such	
Giblet, Gullet & Weasand Meat	Sausage	
Melt	Sold as Such	

SUMMARY	Pounds of Product in Average Hog Weighing 250	Percentage of Green Product to Live Weight of Hog
1. Pork		
Regular Hams	35.00	14.00
Bellies	27.50	11.00
Loins	23.75	9.50
Picnics	13.75	5.50
Boston Butts	11.87	4.75
Fat Backs	7.50	3.00
Lean Trimmings	7.50	3.00
Jowl Butts	5.00	2.00
Spare Ribs	3.75	1.50
Rendered Lard (including leaf)	32.50	13.00
Miscellaneous (including fancy meats from the head)	12.50	5.00
Shrinkage	54.38	21.75
	235.00	94.00
2. Edible By-Products		
Liver, Kidney, Casings, Giblet, Gullet and Weasand Meat, Etc.	8.75	3.50
3. Inedible By-Products		
Grease, Hair, Blood, Tankage, Etc.	6.25	2.50
Total Hog	250.00	100.00

Table 3
Lamb By-Products

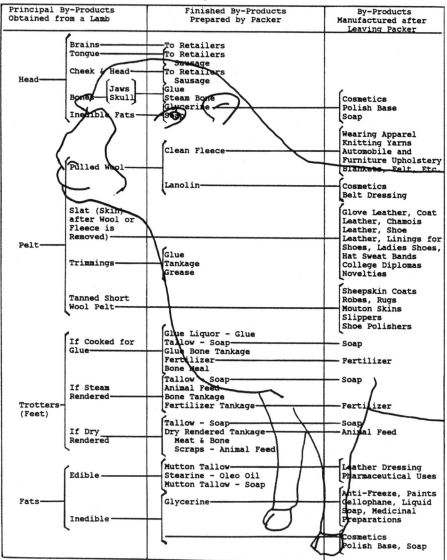

Principal By-Products Obtained from a Lamb	Finished By-Products Prepared by Packer	By-Products Manufactured after Leaving Packer
Head		
Brains	To Retailers	
Tongue	To Retailers Sausage	
Cheek & Head	To Retailers Sausage	
Bones — Jaws Skull	Glue Steam Bone	Cosmetics Polish Base Soap
Inedible Fats	Glycerine 96%	
Pelt		
Pulled Wool	Clean Fleece	Wearing Apparel Knitting Yarns Automobile and Furniture Upholstery Blankets, Felt, Etc.
	Lanolin	Cosmetics Belt Dressing
Slat (Skin, after Wool or Fleece is Removed)		Glove Leather, Coat Leather, Chamois Leather, Shoe Leather, Linings for Shoes, Ladies Shoes, Hat Sweat Bands College Diplomas Novelties
Trimmings	Glue Tankage Grease	
Tanned Short Wool Pelt		Sheepskin Coats Robes, Rugs Mouton Skins Slippers Shoe Polishers
Trotters (Feet)		
If Cooked for Glue	Glue Liquor - Glue Tallow - Soap Glue Bone Tankage Fertilizer Bone Meal	Soap Fertilizer
If Steam Rendered	Tallow - Soap Animal Feed Bone Tankage Fertilizer Tankage	Soap Fertilizer
If Dry Rendered	Tallow - Soap Dry Rendered Tankage Meat & Bone Scraps - Animal Feed	Soap Animal Feed
Fats		
Edible	Mutton Tallow Stearine - Oleo Oil Mutton Tallow - Soap	Leather Dressing Pharmaceutical Uses
Inedible	Glycerine	Anti-Freeze, Paints Cellophane, Liquid Soap, Medicinal Preparations Cosmetics Polish Base, Soap

Source: Swift and company, c. 1970

Principal By-Products Obtained from a Lamb	Finished By-Products Prepared by Packer	By-Products Manufactured after Packer

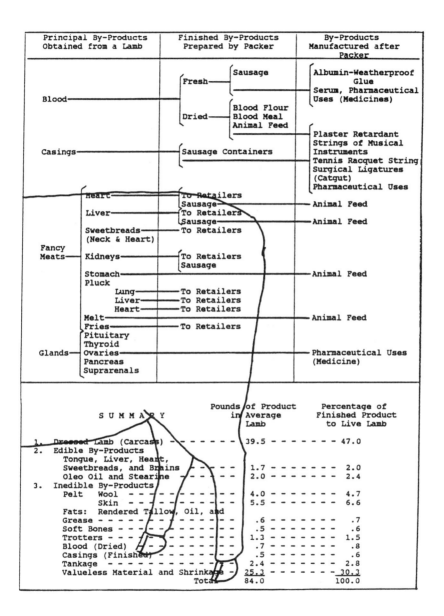

Blood —
  Fresh — Sausage
  Dried — Blood Flour / Blood Meal / Animal Feed

Albumin-Weatherproof Glue
Serum, Pharmaceutical Uses (Medicines)

Casings — Sausage Containers

Plaster Retardant
Strings of Musical Instruments
Tennis Racquet String
Surgical Ligatures (Catgut)
Pharmaceutical Uses

Fancy Meats —
  Heart — To Retailers / Sausage — Animal Feed
  Liver — To Retailers / Sausage — Animal Feed
  Sweetbreads (Neck & Heart) — To Retailers
  Kidneys — To Retailers / Sausage
  Stomach — Animal Feed
  Pluck
    Lung — To Retailers
    Liver — To Retailers
    Heart — To Retailers
  Melt — Animal Feed
  Fries — To Retailers

Glands —
  Pituitary
  Thyroid
  Ovaries — Pharmaceutical Uses (Medicine)
  Pancreas
  Suprarenals

S U M M A R Y	Pounds of Product in Average Lamb	Percentage of Finished Product to Live Lamb
1. Dressed Lamb (Carcass)	39.5	47.0
2. Edible By-Products		
Tongue, Liver, Heart, Sweetbreads, and Brains	1.7	2.0
Oleo Oil and Stearine	2.0	2.4
3. Inedible By-Products		
Pelt Wool	4.0	4.7
Skin	5.5	6.6
Fats: Rendered Tallow, Oil, and Grease	.6	.7
Soft Bones	.5	.6
Trotters	1.3	1.5
Blood (Dried)	.7	.8
Casings (Finished)	.5	.6
Tankage	2.4	2.8
Valueless Material and Shrinkage	25.3	30.3
Total	84.0	100.0

Table 4
Value[a] of Plant Nutrients in Different
Sources of Animal Excreta

Nutrient	Poultry Litter	Dehydrated Poultry Excreta	Cattle Excreta	Swine Excreta
Nitrogen	12.62	11.29	8.19	9.48
Potassium	3.77	4.93	1.06	2.84
Phosphorus	16.02	22.25	14.24	18.96
Calcium	.26	.95	.09	.29
Total plant nutrient value	32.67	39.42	23.58	31.57

Source: Smith and Wheeler, 1979.

[a]$/1000 kg dry matter. Based upon prices paid by farmers cost of N-$.25/kg from anhydrous ammonia; K-$.21/kg from muriate of potash; P-$.89/kg from super phosphate; Ca-$.01/kg from ground limestone. Ground limestone price at quarry.

manure being incorporated into the animal feed was addressed in ARS publication 44-224, authored by Smith et al. (1971). The challenge of feeding manure to animals without adversely influencing animal health can be met by subjecting the manure to high temperatures for an adequate time to destroy bacteria and other microorganisms. Shortcuts may result in the transmission of disease to healthy slaughter animals. This concern is particularly acute as a result of scientists identifying not only meat animals as a source of bacteria pathogenic to humans but also the animal-to-man transmission of antimicrobial-resistant bacteria like salmonellae (Holmberg et al., 1984). Very recently, Larkin (1985) indicated

Table 5
Nutrient Composition of Animal Excreta[a]

Item	Poultry Litter	Dehydrated Poultry Excreta	Cattle Excreta (Steer)	Swine Excreta
Total digestible nutrients	73.0	52.0	48.0	48.0
Crude protein (N × 6.25)	31.0	28.0	20.0	24.0
Crude fiber	17.0	13.0	20.0	15.0
Calcium	2.4	8.8	0.9	2.7
Phosphorus	1.8	2.5	1.6	2.1
Magnesium	0.4	0.7	0.4	0.9
Potassium	1.8	2.3	0.5	1.3

Source: Smith and Wheeler, 1979.

[a]Percentage of dry matter.

Table 6
Estimated Daily Dry Matter in Excreta
from Selected Classes of Farm Animals

Class	Live Body Weight (kg)	Estimated Intake (kg)	Daily Feed Dry Matter Digestibility (%)	Excreta Dry Matter/Day[a] (kg)
Poultry				
Laying hens	1.8	0.110	80	0.022
Broilers	0.75	0.075	85	0.011
Swine (growing)	40.0	2.5	77	0.58
Ruminants				
Growing beef steers	200	4.9	60	1.96
Finishing beef steers	400	10.3	80	2.06
Dry pregnant mature beef cows	550	17.6	50	8.80
Sheep (finishing)	35	1.4	67	0.46
Growing dairy heifers	250	6.5	62	2.47
Lactating dairy cows (large breeds, 40 kg milk, daily)	550	22.0	75	5.50

Source: Smith, 1977.

[a]Estimates do not include nutrients excreted in urine, except for poultry.

that animal fecal material is "loaded" with viruses and therefore the indiscriminate use of manure in feed should be discouraged.

Animal manure from the meat establishment may also be utilized to produce methane as a substitute for propane or natural gas. Patel (1966) described the method of producing methane from manure that has been used in India for decades. He indicated that 1 cubic m of methane gas is equivalent to 1.5 cubic m of manufactured gas or 2.2 kw of electrical energy. The digestion is about 12 times as efficient at 35° C than at 15° C. The presence of antibiotic residues in the manure interferes with the fermentation process. The carbon dioxide which is produced during the normal fermentation can be absorbed in water and the hydrogen sulfide can be removed by passing the methane gas through a suspension of ferric oxide. The pure form of methane that results can be used in the laboratory for bunsen burners and, more importantly, in the meat establishment as a replacement for natural gas.

Patel's calculations suggest the cost of a small digester can be paid for by savings in fuel in about 8 mo. Varel *et al.* (1980) reported about 1.6 l of methane can be produced per liter of fermenter at 35° C. The effluent from the digester retains its fertilizing capacity but does not have an

Table 7
Amino Acid Content of Farm Animal Feces

Amino Acids	Beef Cattle	Sheep	Swine	Poultry
	Amino Acid g/16 g N[a]			
Arginine	1.1	3.4	3.1	4.2
Histidine	0.7	1.3	1.8	1.8
Isoleucine	1.3	3.7	4.8	4.5
Leucine	3.8	4.4	7.2	7.3
Methionine	2.9	5.1	5.1	4.4
Phenylalanine	0.6	1.5	2.7	0.8
Threonine	0.0	3.4	4.0	4.0
Tryptophan	1.8	4.8	3.7	4.5
Valine	—	0.9	—	—
Aspartic acid	2.3	4.9	4.8	5.7
Serine	4.3	—	6.3	9.7
Glutamic acid	1.5	—	2.7	4.7
Proline	3.8	—	15.6	14.0
Glycine	1.8	—	4.2	4.9
Alanine	2.7	—	7.0	7.5
Cystine	4.0	—	5.3	9.7
Tyrosine	0.2	—	3.0	2.3
Total amino acids (g)	32.5	33.4	81.9	100.0

Source: Smith, 1973.
[a] g amino acid/16 g true fecal protein nitrogen.

odor. An effective method of moving manure was described by Miller (1983). The system, which is presently in use on at least one farm, utilizes compressed air to move the manure from the collecting area to the holding area. The manure could be moved just as readily into the digester using this method.

Dressing Percentages

Once the livestock have been slaughtered, the animal parts are segregated on the basis of end use suitability. As described by Breidenstein later in this book, a relatively small portion of the animal is ultimately used for human consumption. In beef, the dressing percentage, which is the weight of the carcass compared to the animal live weight, can be estimated to be about 60%. That average varies depending on the type of cattle being slaughtered (Table 8; USDA, 1969). Table 8 also provides average dressing percentages for calves/veal, lambs/sheep, and barrows/gilts. By-products as discussed in this section of the book will

Table 8
Dressing Percentages of Various Kinds of Livestock, by Grades[a]

Livestock Species/Grade	Range	Average
Cattle		
Prime	62–67	64
Choice	59–65	62
Good	58–62	60
Standard	55–60	57
Commercial	54–62	57
Utility	49–57	53
Cutter	45–54	49
Canner	40–48	45
Calves/Vealers		
Prime	62–67	64
Choice	58–64	60
Good	56–60	58
Standard	52–57	55
Utility	47–54	51
Lambs/Sheep		
Prime	49–55	52
Choice	47–52	50
Good	45–49	47
Utility	43–47	45
Cull	40–45	42
Barrows/Gilts		
U.S. No. 1	68–72	70
U.S. No. 2	69–73	71
U.S. No. 3	70–74	72
U.S. No. 4	71–75	73
Utility	67–71	69

Source: USDA, AMS-Livestock Division.
[a]All percentages are based on hot weights.

focus on the 30–50% of the live animal that is not part of the carcass. However, some attention will also be given to bones because of the recent growing interest in stripping muscle groups from the bones before the carcass has been chilled (Cross, 1980).

Bengtsson and Holmquist (1984) reported the approximate composition of cattle to be: carcass (meat and bone) 50%; organs 16%; hide 6%; blood 3%; fat tissue 4%; horns, skull, and feet 5%; and abdominal and intestinal contents 16%. They listed the biologically and hygienically edible parts to be: meat (bones removed) 34%; organs 14%; blood 3%; and fat tissue 4%; for a total of 55% of the live animal weight. A greater percentage of meat is obtained from swine. Those authors reported the

Table 9
Dressing Percentages of Chilled
Whole Eviscerated Poultry[a]

| | % Yield[b] | | |
Class of Poultry	Minimum	Maximum	Average
Chickens			
Broilers, fryers	70	75	73
Roasters	72	76	74
Fowl	70	76	73
Turkeys			
Fryer-roasters	75	81	79
Light breeds	76	83	79
Heavy hens	78	84	81
Heavy toms	78	84	82
Ducks			
Ducklings (7–9 weeks)	70	75	73
Geese			
White Chinese (6–8 weeks)	—	—	68
White Chinese (10–16 weeks)	72	75	73
White Chinese (6–8 mo)	—	—	73
Heavy breeds (5–8 mo)	65	71	68
Heavy breeds (10–16 weeks)	68	75	73
Heavy breeds (over 16 weeks)	70	75	73

Source: Swanson *et al.*, 1964.
[a]Including neck and giblets. [b]Based on live weight.

approximate composition of swine to be: carcass (meat and bone) 69%; organs 7%; hide 6%; blood 3%; fat tissue 3%; hooves and skull 6%; and abdominal and intestinal contents 6%. The edible portions were: meat (bones removed) 52%; organs 5%; blood 3%; and fat tissue 3%; for a total of 63% of the live animal weight as edible.

The dressing percentages for poultry range between 70 and 84%, as presented in Table 9 (Swanson *et al.*, 1964). A raw turkey carcass of 6.5 to 7.3 kg will consist of 59.3% meat, 26.9% breast, 14.0% thigh, 31.1% bone, and 7.5% skin, whereas when the carcass weight increases between 13.0 and 14.4 kg, it will consist of 60.8% meat, 29.5% breast, 13.2% thigh, 25% bone, and 11.9% skin (Carlson *et al.*, 1975). Table 10 (Winter and Clements, 1957) provides the percentage distribution of cut-up parts in different species of ready-to-cook poultry. Tables 11 and 12 present the weight of the inedible portion of chickens and turkeys, such as feathers, lungs, etc. These inedible products do have value as rendered products to be used as feed supplements. For example, if feathers are rendered in a closed container with saturated steam under 40 to 60

Table 10
Percentage Distribution of Cut-Up Parts in Different Species of Ready-to-Cook Poultry

Species	Number of Birds	Average Age (Weeks)	Average Live Weight (lb)	Legs and Thighs (%)	Breast (%)	Back and Ribs (%)	Wings (%)	Necks (%)	Gizzard (%)	Liver (%)	Heart (%)
Broilers											
Male	10	10.0	3.9	34.1	25.4	17.0	13.3	3.5	3.5	2.6	0.6
Female	10	10.0	3.0	32.4	25.7	16.6	13.6	4.1	4.1	2.8	0.7
Turkeys											
Small											
Male	10	16.0	10.0	29.3	28.3	19.9	13.0	3.3	3.8	2.3	1.0
Female	10	16.0	7.0	29.0	29.9	20.0	12.0	3.7	3.8	2.1	0.4
Large											
Male	10	28.0	22.0	24.1	41.3	18.4	9.5	3.4	1.8	1.1	0.4
Female	10	28.0	14.0	25.5	35.5	20.6	11.3	2.4	2.8	1.6	0.3
Ducks											
Male	10	7.5	7.1	23.4	29.7	23.0	10.6	5.4	4.3	2.7	0.9
Female	10	7.5	6.3	23.5	30.1	23.6	10.8	4.7	3.9	2.5	0.9
Geese	20	10–12	10.8	21.9	23.8	21.3	16.0	6.3	6.3	3.5	0.9

Source: Winter and Clements, 1957.

Table 11
Average Weights of Parts of Chicken Carcass

| | Live Slaughter Weight | | |
	3 lb	4 lb	5 lb
Age in days	94	189	219
Percentage "fill"	3.7	2.6	3.7
Empty weight (g)	1293	1787	2245
Offal	(%)	(%)	(%)
Feathers	8.4	8.5	7.5
Blood	3.2	3.3	3.5
Head	2.9	2.7	2.4
Shanks and feet	4.7	3.7	3.6
Total offal	19.1	18.2	16.9
Viscera			
Heart	0.42	0.48	0.45
Liver	2.2	1.7	1.9
Kidneys	0.63	0.55	0.62
Pancreas	0.22	0.24	0.22
Spleen	0.23	0.18	0.21
Lungs	0.51	0.45	0.40
Digestive tract	10.1	9.4	8.6
Total viscera	14.2	12.9	12.4
Dressed Carcass			
Skin	8.0	9.2	10.0
Neck	3.5	2.9	2.7
Legs above hock	20.0	19.3	19.0
Wings	6.3	5.4	5.4
Torso	26.7	29.3	30.2
Total dressed carcass	64.4	66.1	67.3
Total bone in dressed carcass	17.3	15.0	14.7
Total flesh and fat in dressed carcass	38.4	41.0	41.8
Total flesh, fat, and edible viscera[a]	45.3	47.0	47.6

[a]Including heart, liver, gizzard, and kidneys.

P.S.I. for about an hour, a product is obtained which after drying will have a good bulk density and good digestibility when mixed with other feed ingredients. Commercial feather meal will contain about 82–90% crude protein, 2.5–4.0% crude fat, 0.1–0.4% crude fiber, 2.1–4.2% ash, 5.6–11.8% water, and 0.1–0.4% nitrogen-free extract. Table 13 (Panda and Sharma, 1974) provides the nutrient composition of hydrolyzed feather meal, poultry blood meal, dehydrated poultry manure, and poultry by-product meal.

Table 12
Average Weights of Parts of Turkey Carcass

	Male (22.6 lb)	Female (21.5 lb)
Live starved weight	100.0%	100.0%
Dressed and chilled (New York)	90.76	91.15
Blood and feathers (by diff.)	9.24	8.85
Shanks and feet	2.92	2.51
Head	1.68	1.65
Intestine, caeca, and cloaca	2.35	2.62
Esophagus, crop, and proventriculus	0.80	0.73
Lungs	0.29	0.27
Trachea	0.16	0.15
Pancreas	0.12	0.16
Spleen	0.06	0.04
Total viscera excluding giblets	3.78	3.97
Liver	1.12	1.03
Heart	0.36	0.33
Gizzard	1.91	1.95
Total including giblets	3.39	3.31
Kidneys	0.26	0.29
Carcass without giblets	73.50	72.09

Blood

The blood recovered from slaughter animals in the United States is used for inedible purposes such as fertilizer or feed ingredients. For such uses, it can be rendered with other offal or dried (Bengtsson and Holmquist, 1984). The yield of blood has been reported to be 10–12 l from beef, 2.5 l from hogs, and 1.5 l from sheep (Wismer-Pedersen, 1979). Skrzynski (1962) indicated blood recoveries of 5–17 kg from horses, 7–14 kg from cattle, and 1–2 kg from calves and sheep, under commercial conditions. As seen in Fig. 1 (Kotula and Helbacka, 1966), the amount of blood in chickens ranges between 7 and 12% of their body weight, depending on their weight (Newell and Shaffner, 1950). During slaughter, the birds can be expected to lose between 44 and 54% of their blood (Kotula and Helbacka, 1966).

Until recently, blood was routinely emptied into the sewage lines (*Meat Industry,* 1978). The cost of treating the blood of one beef animal as sewage in the municipal sewage treatment plant was estimated to be

Table 13
Nutrient Composition of Inedible By-Products

Ingredients	Dehydrated Poultry Manure	Poultry By-Product Meal	Hydrolyzed Feather Meal	Poultry By-Product and Feather Meal Combined	Poultry Blood Meal	Hatchery By-Product Meal
Crude protein (%)	27.0	54.5	81.5	55.9	72.0	26.0
Ether extract (%)	1.8	14.6	6.7	21.7	3.1	11.4
ME Kcal/Kg	900	2860	2354	3610	2685[a]	1694
Ca (%)	7.4	4.52	0.66	3.9	1.50	20.60
P (%)	2.1	2.23	0.74	0.8	0.64	0.49
Arginine (%)	0.41	5.9	5.8	0.43	5.2	4.8
Glycine (%)	0.51	10.9	7.0	4.87	4.4	5.5
Histidine (%)	1.19	1.5	0.8	1.13	4.1	2.0
Isoleucine (%)	0.37	3.2	4.1	2.57	3.6	3.6
Leucine (%)	0.60	5.6	6.7	4.38	8.6	6.1
Lysine (%)	0.39	4.5	2.4	2.42	6.8	4.1
Methionine (%)	0.12	1.4	0.7	0.61	1.2	1.9
Cystine (%)	0.21	.	3.4	3.19	0.8	1.1
Phenylanine (%)	0.39	3.0	3.8	2.60	4.8	3.5
Tyrosine (%)	0.27	1.7	1.3	1.71	2.1	2.3
Threonine (%)	0.40	2.9	3.5	2.61	3.9	3.4
Tryptophan (%)	0.53	0.7	0.7	—	1.9	1.3
Valine (%)	0.52	4.3	6.7	3.66	6.3	5.0

Source: Panda and Sharma, 1974.
[a]Approximated from comparable organic constituents in poultry by-product meal.

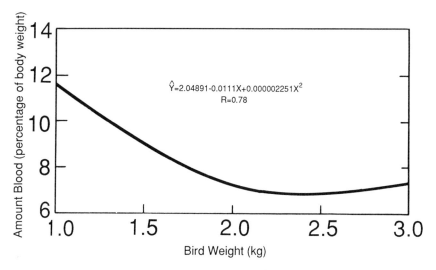

Figure 1   Blood from chickens as related to body weight. (Kotula, 1966).

equivalent to the sewage from eight homes. Such calculations quickly underscored the benefits of collecting the blood and either separating the cells from the plasma by centrifugation or by drying the whole blood for use in feed or fertilizers. In many European countries, animal blood is utilized in sausage, soups, and other foods. During bleeding, the blood is collected by use of a hollow knife fastened to a vacuum system (Wismer-Pedersen, 1979). The blood from several animals is combined and retained until each of the carcasses has successfully passed inspection.

Blood from slaughter animals contains about 18% of high-quality protein (USDA, 1984). The strong flavor and aroma associated with dried plasma and hemoglobin identified by West and Young (1983) can probably be minimized by low temperature drying methods (Stevenson and Lloyd, 1979). Dill (1976), Sato *et al.* (1981), and Quaglia and Massauri (1982) described the separation of blood proteins. The flow diagram in Fig. 2 (Dill, 1976) shows the procedures for the preparation of protein isolates from blood. Dill reported that when up to 10% of bread flour was replaced by plasma isolate, the loaf volume increased 16–18% and the essential amino acid content of the resultant bread was increased significantly. Of particular importance is the high level of lysine, about 9% in blood, because that is sometimes the limiting protein in non-animal foods (Waibel, 1974). Heat treatment to destroy possible pathogenic microorganisms in the blood may also cause a rather extensive destruction of lysine (Danish Meat Research Institute, 1963). The

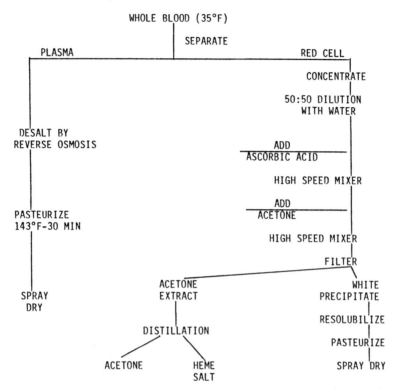

WHOLE BLOOD (35°F)

SEPARATE

PLASMA

RED CELL

CONCENTRATE

50:50 DILUTION
WITH WATER

DESALT BY
REVERSE OSMOSIS

ADD
ASCORBIC ACID

HIGH SPEED MIXER

ADD
ACETONE

PASTEURIZE
143°F-30 MIN

HIGH SPEED MIXER

FILTER

ACETONE
EXTRACT

WHITE
PRECIPITATE

SPRAY
DRY

RESOLUBILIZE

DISTILLATION

PASTEURIZE

ACETONE

HEME
SALT

SPRAY DRY

Figure 2 Flow diagram for preparation of protein isolates from blood (Dill, 1976).

amounts of the other amino acids in blood meal are shown in Table 14 (Waibel, 1974). Blood is also an excellent source of heme iron (Slinde *et al.*, 1982), therefore its addition to food would be nutritionally beneficial. Slinde *et al.* indicated the addition of 1–2.5% whole blood to sausage improved the color, flavor, and meat taste. Akers (1973) reported the baking industry might be interested in the use of plasma as a substitute for egg albumen, to reduce costs. Braathen (1982) reported blood plasma can be fried to produce a product similar to scrambled eggs.

Hides and Skins

Most unmarred cattle hides are used in the manufacture of leather (USDA, 1976; Feairheller, 1977). The standards and terms used for patterning, selection, cure, fleshing, and tare allowances in packing are presented in "Trade Practice for Proper Packer Cattle Hide Delivery,"

Table 14
Percentage Amino Acid Composition
of Blood Meals

	B–W[b]	S–N–Y[c]	University of Minnesota[a]	
			Vat[d]	Ring[e]
N × 6.25	84.0	80.0	82.1	89.6
Arginine	3.3	3.5	3.6	3.8
Histidine	5.0	4.2	3.5	4.4
Lysine	6.7	6.9	7.0	8.5
Tryptophan	1.1	1.1		
Phenylalanine	6.0	6.1	5.7	6.5
Cystine	1.5	1.4		
Methionine	1.0	0.9	1.0	1.2
Threonine	3.8	3.7	3.1	4.0
Leucine	13.9	10.3	10.7	12.0
Isoleucine	0.95	1.0	0.96	0.94
Valine	8.25	6.5	7.4	8.7
Glycine		3.4	4.6	4.0
Tyrosine		1.8	2.1	2.6
Aspartic acid			9.1	10.0
Serine			3.1	3.9
Glutamic acid			8.2	8.8
Proline			3.8	3.7
Alanine			6.8	7.4
Dry matter			93.6	91.4

Source: Waibel, 1974.
[a]Block, R. J., and Weiss, K. W. (1956). "Amino Acid Handbook." Charles C. Thrones, Springfield, Illinois. [b]Scott, M. L., Nesheim, M. C., and Young, R. J. (1969). "Nutrition of the Chicken." Scott and Associates, Ithaca, New York. [c]Analyses conducted using ion exchange chromatography at the University of Minnesota under FPRF grant. [d]Refers to conventional process. Average of 12 samples reported. [e]Refers to ring-dried blood. Average of 6 samples reported.

published by the Skin and Leather Association and the Leather Industries of America (*The National Provisioner*, 1985). Poor-quality hides and trimmings can be used to produce products such as dog chews, gelatin, and sausage casings (Lipsett, 1980). The value of dog chews in 1980 was 8 cents per pound and gelatin was 6 cents per pound, whereas sausage casings were 75 cents per pound. Hide prices, at that time, ranged between 45 and 73 cents per pound. The procedure for preparing sausage casings from hides is considered proprietary, due in part to the high value placed on such products in comparison to gelatin and dog chews

(Lipsett, 1980). By-products of the tanner industry, including chrome-tanned leather material such as blue trimmings, splits, and shavings as well as salted or cured hide trimmings and fleshings, are used for the production of animal glue, fertilizer, leather and fiberboard, protein hydrolysate, and collagen or as landfill (Pearson, 1982). The skinning of hogs has been reported to be economically feasible (Judge *et al.*, 1978; Kaiser, 1980).

Paunch and Intestinal Contents

The average wet weight of the paunch contents of cattle, sheep, and lambs are 27, 2.7, and 1.7 kg, respectively (Fernando, 1980). The biological oxygen demand (BOD) (mg/l) is 50,000 for cattle and 30,000 for sheep, so since 80% of the BOD of paunch contents is soluble, its disposal poses a serious concern. Fernando proposed separating the liquid fraction from the fibrous residue by a decanter centrifuge or a press and then using ultrafiltration to concentrate the protein in the paunch contents. The fibrous fraction could be used as fuel and the concentrate as pig feed. The amino acid composition of paunch content material concentrate in g/100 g protein was reported by Fernando as lysine, 4.7; histidine, 1.2; arginine, 2.9; aspartic acid, 7.3; threonine, 3.9; serine, 3.2; glutamic acid, 9.1; proline, 2.9; glycine, 4.2; alanine, 6.4; half cystine, 0.7; valine, 4.6; methionine, 2.3; isoleucine, 4.3; leucine, 7.1; tyrosine, 6.2; phenylalanine, 3.7; and tryptophan, 0.8. The amino acid composition of the concentrate suggests the concentrate might be substituted successfully for cheese whey in pig feed rations. The use of this system is anticipated to reduce the cost of disposal of paunch contents by about 50%, according to the author of the study. Shih and Kozink (1980) reported a similar use of ultrafiltration of poultry processing waste water to recover a nutritional by-product containing 30–35% protein and 24–45% fat. The authors suggest the chemical score based on amino acid content was similar to soybean protein.

Bone

With the continued interest in hot boning—the removal of muscle groups from the intact bones of the carcass before chilling (Cross, 1980)—the availability of bones at the meat establishments will increase dramatically. Selected bone groups will be crushed and sent through equipment to remove adhering meat and bone marrow to produce mechanically separated meat. Wismer-Pedersen (1979) indicated the average yield of mechanically separated meat from normally trimmed beef bones was 25% and 35% for port bones. The composition of mechanically separated beef was: protein 15%, fat 25%, moisture 58%, and ash 1–2%; whereas for pork, it was: protein 13%, fat 28%, moisture 58%, and ash 1–3%, according to Wismer-Pedersen. The bone and connective tissue fragment resulting from the mechanical separation process can be further treated to obtain fat, edible bone collagen, edible bone phosphate, and edible bone protein (Jackson *et al.*, 1982; USDA, 1984). Jobling and Jobling (1983) described the defatting process, the acid process, and the cooking process to obtain these products from bone. Such products are not yet permitted in the United States but are being produced in some European countries. A generalized flow diagram for the production of bone extract was presented by Ochi (1981). Such bone extract has been used effectively for many years to add flavor to soups, barbecue sauces, retorted stews, curry, hams, fruit pastes, and sausages.

References

Akers, J. M. (1973). Utilization of Blood. *Food Manufacture* **48(4)**, 31–32.

Animal blood protein as a food ingredient. Memorandum of screening and surveillance, FSIS, *USDA* **3(1)**, 5–7, 1984.

Areas, J. A. G., and Lawrie, R. A. (1984). Effect of lipid-protein interactions on extrusion of offal protein isolates. *Meat Science* **11**, 275–299.

Bengtsson, O., and Holmquist, O. (1984). By-products from slaughtering. *Fleischwirtschaft* **64(3)**, 334–336.

Blood dryer solves costly sewage problem. *Meat Industry,* November 1978, p. 42.

Blood meal. Danish Meat Research Institute, Publication No. 40, p. 18, 1963.

Braathen, O. S., Nilsen, J. A., and McCormick, R. (1982). Ideas for the use of animal blood plasma. Proceedings of the 28th European Meeting of Meat Research Workers, p. 344.

Carlson, C. W., Marion, W. W., Miller, B. F., and Goodwin, T. L. (1975). Factors affecting poultry meat yields. North Central Regional Research Publication No. 226.

Copeland, K. (1982). Poultry litter makes cheap beef ration. *Progr. Farmer,* **March,** 13–19.

Cross, H. R. (1980). Optimal systems for rapid processing of beef. Proceedings of the 26th European Meeting of the Meat Research Workers, Section H-4, pp. 15–18. Colorado Springs, Colorado.

Dailey, J. P. (1977). Pharmaceuticals and medicinals of animal origin. Proceedings of the Meat Industry Research Conference, American Meat Science Association, pp. 87–92.

Dill, C. W. (1976). Use of plasma in edible meat products. Proceedings of the 29th Annual Reciprocal Meat Conference, American Meat Science Association, pp. 162–177.

Edible bone protein. Memorandum of screening and surveillance, FSIS, *USDA* **3(3)**, 25–26, 1984.

Feairheller, S. H. (1977). Future trends in hide handling. Proceedings of the Meat Industry Research Conference, American Meat Science Association, pp. 93–96.

Fernando, T. (1980). Utilization of (cattle) paunch content material by ultrafiltration. *Process Biochem.* **15(3)**, 7–9.

Forrest, J. C., Aberle, E. D., Hedrick, H. B., Judge, M. D., and Merkel, R. A. (1975). "Principles of Meat Science." W. H. Freeman and Co., San Francisco.

Holmberg, S. D., Wells, J. G., and Cohen, M. L. (1984). Animal-to-man transmission of antimicrobial resistant *Salmonella:* Investigation of U.S. outbreaks, 1971–1983. *Science,* **August 24,** 833–835.

Jackson, E. D., Consolacion, F. I., and Jelen, P. (1982). Bacteriological evaluation of alkali-extracted protein from poultry residues. *J. Food Protection* **45(9),** 797–800.

Jobling, A., and Jobling, C. A. (1983). Conversion of bone to edible products. Proceedings **35,** p. 93. Easter School in Agricultural Science, University of Nottingham.

Judge, M. D., Salm, C. P., and Okos, M. R. (1978). Hog skinning versus scalding. Proceedings of the Meat Industry Research Conference, American Meat Science Association, pp. 155–164.

Kaiser, W. (1980). Hog skinning pays: Here's how it's done at Kayser's in Lena, Ill. *National Provisioner,* September 13, pp. 44–46.

Kotula, A. W., and Helbacka, N. V. (1966). Blood volume of live chickens and influences of slaughter technique on blood loss. *Poult. Sci.* **45(4),** 684–688.

Larkin, E. P. (1985). Detection and quantitation of foodborne viruses. Proceedings basic symposium "Foodborne Microorganisms and Their Toxins." Institute of Food Technologists (Abstract).

Levie, A. (1963). "The Meat Handbook." Avi Publishing Co., Inc., Westport, Connecticut.

Levin, E. (1970). Conversion of meat by-products into edible meat protein concentrate powder. Proceedings of the Meat Industry Research Conference, American Meat Science Association, pp. 29–38.

Lipsett, V. A. (1980). Collagen utilization. *J. Amer. Leather Chemists Assn.* **75(11),** 439–455.

Manure found as cheap fertilizer source. *Feedstuffs* **(57)18,** 18, 1985.

Miller, V. (1983). Now they move manure by air. *Prog. Farmer,* **March,** C-1, 2.

More than meat. *USDA* **35(2),** 3, 1976.

Newell, G. W., and Shaffner, C. S. (1950). Blood volume determination in chickens. *Poult. Sci.* **29,** 78–87.

New trade practices book on cattle hides issued. *National Provisioner,* May 25, 1985, pp. 32–34.

Ochi, H. (1981). Production and applications of bone extracts. *Food Technology* **35,** 58–59.

Olson, F. C. (1970). Nutritional aspects of offal proteins. Proceedings of the Meat Industry Research Conference, American Meat Science Association, pp. 23–28.

Panda, B., and Sharma, N. (1974). Inedible by-products of poultry industry: A review. *Poult. Guide* **11(6),** 19–26.

Patel, J. D. (1966). Digester converts poultry manure into useful methane gas. *Poultry Tribune* **72(12),** 31.

Pearson, C. L. (1982). Animal glue in industry and the use of chrome-tanned hide collagen in the production of animal glue. *J. Amer. Leather Chemists Assn.* **77(6),** 306–310.

Quaglia, G. B., and Massauri, A. (1982). Proteolysates from slaughterhouse blood. *J. Sci. Food Agric.* **33,** 634.

Romans, J. R., and Ziegler, P. T. (1974). "The Meat We Eat." Interstate Printers and Publishers, Inc., Danville, Illinois.

Sato, Y., Hoyakawa, S., and Hoyakawa, M. (1981). Preparation of blood globin through carboxymethyl cellulose chromatography. *J. Food Technology* **16,** 81.

Shih, J. C. H., and Kozink, M. B. (1980). Ultrafiltration treatment of poultry processing wastewater and recovery of a nutritional by-product. *Poult. Sci.* **59(2),** 247–252.

Skrzynski, T. (1962). Zbiorka krwi w zakloduck miesnych. *Gospodarka Miesna,* **8,** 14, 15.

Slinde, E. Mielnik, J., Martens, M., and Tenningen, A. (1982). Utilization of animal blood for human consumption: Sensory evaluation and measurements of surface colour of meat products and determination of hemoglobin and myoglobin in raw materials. Proceedings of the 28th European Meeting of Meat Research Workers, p. 340.

Smith, L. W. (1973). Recycling animal wastes as protein sources: Alternative sources of protein for animal production. Sym. Proc. National Academy of Science pp. 146–173.

Smith, L. W. (1977). The nutritional potential of recycled wastes. New feed sources, Proc. tech. consultation. Food and Agriculture Organization of the United Nations, Rome, pp. 227–244.

Smith, L. W., and Wheeler, W. E. (1979). Nutritional and economic value of animal excreta. *J. Anim. Sci.* **48(1),** 14–156.

Smith, L. W., Calvert, C. C., Frobish, L. T., Dinius, D. A., and Miller, R. W. (1971). Animal waste reuse—Nutritive value and potential problems from feed additives: A review. USDA, ARS publication 44–224.

Stevenson, T. R., and Lloyd, C. J. (1979). Better uses for abattoir blood. *Ag. Gazette of New So. Wales,* **90,** 42.

Swanson, M. H., Carlson, C. W., and Fry, J. L. (1964). Factors affecting poultry meat yields. North Central Regional Research Publication No. 158.

Swift Independent Porkers (c. 1950). Cattle by-products. Swift Agriculture Bulletin 6, Chicago.

Swift Independent Porkers (c. 1955). Hog by-products. Swift Agriculture Bulletin 10, Chicago.

Swift Independent Porkers (c. 1970). Lamb by-products. Swift Agriculture Bulletin 19, Chicago.

U.S. Department of Agriculture (1969). Approximate average and range in dressing percentages of various kinds of livestock by grades. USDA, Consumer and Marketing Service.

Varel, V. H., Hashimoto, A. G., and Chen, Y. R. (1980). Effect of temperature and retention time on methane production from beef cattle waste. *Appl. and Envir. Microb.* **40(2),** 217–222.

Waibel, P. F. (1974). Processing and nutritional value of blood meals. *Meat Processing* **13(5),** 102–103.

West, S. I., and Young, J. N. (1983). The protein alternatives. *Food Manufacture,* **March,** 22.

Winter, A. R., and Clements, P. (1957). Cooked edible meat in ready-to-cook poultry. *J. Amer. Diet. Assn.* **33(8),** 800–802.

Wismer-Pedersen, J. (1979). Utilization of animal blood in meat products. *J. Food Technol.* **33,** 76.

Contribution of Red Meat to the U.S. Diet

B. C. Breidenstein

National Live Stock and Meat Board (Retired)
Chicago, Illinois

and

J. C. Williams

National Dairy Promotion
and Research Board
Arlington, Virginia

Handbook of Animal Science

Executive Summary

Considerable confusion exists with regard to the amount of red meat actually ingested in the United States. For purposes of following trends or general changes in ingestion, annual per capita carcass weight produced is a useful number and has the merit of long-term continuity. That per capita production is often called "consumption." It is, of course, in the economic sense in that all of that material was consumed. Confusion arises, however, when "consumption" is presumed to be analogous to "ingestion." The boneless lean mass (usually cooked), with a small amount of adhering fat tissue, represents meat truly ingested. The carcass weight "consumed" contains large amounts of bone and fat, as well as evaporative and drip cooking losses that are not, however, "ingested."

Using beef as the example and following its conversion from carcass to ingestion will serve to demonstrate the difference between "consumption," as historically assessed, and "ingestion." Daily per capita beef "consumption" in carcass weight terms in 1984 was about 131.98 g (4.65 oz), bones represented about 21.55 g (.76 oz), fat tissue for rendering about 7.37 g (.26 oz), and retail trim, cooking losses, and plate waste about 36.27 g (1.27 oz). The beef that was ultimately ingested consisted

Table 1

Cooked Meat Ingestion for All Red Meats in the U.S. Diet, 1984

	Light Users		Moderate Users		Heavy Users		Total Available Annually (Billions of Pounds)
	g	oz	g	oz	g	oz	
Fresh beef	16.31	.57	42.16	1.49	67.44	2.38	7.1870
Ground beef	7.89	.28	17.50	.62	30.14	1.06	3.1679
Fresh pork	2.80	.10	11.17	.39	21.99	.78	2.0140
Fresh lamb	.32	.01	.62	.02	1.23	.04	.1233
Fresh veal	.63	.02	1.22	.04	2.41	.09	.2420
Processed meat	13.18	.47	44.33	1.56	93.11	3.28	8.3581
Total red meat ingested	41.13	1.45	117.00	4.12	216.32	7.63	21.0923

of about 54.50 g (1.92 oz) fresh plus 9.41 g (.33 oz) as processed meat. Daily per capita carcass weight "consumption" was about 131.98 g (4.65 oz), whereas daily per capita ingested beef was about 63.91 g (2.25 oz).

Thus, while per capita carcass weight disappearance is very useful for a number of purposes, it does not provide useful dietary information. Also, it is well recognized that different consumers consume food at different levels. Recent surveys (Yankelovich, Skelly and White, Inc., 1985) permit segmenting the population into different user levels for all species combined, as shown in Table 1.

The varying levels of meat ingestion are frequently debated on the grounds that the number of nonconsumers in most cases remains unknown. Using the population segmentation and use level estimates (Yankelovich, Skelly and White, Inc., 1985) permits one to speculate with a higher degree of accuracy than has previously been possible. When overall red meat ingestion was derived, some 5% of the total population were nonusers of fresh red meat and 9% nonusers of processed red meat. Based on these population numbers, it is possible to determine nutritional contribution to the diet of total red meat, as shown in Table 2.

In estimating dietary contribution of fresh red meats to the diet, the assumption has been made that one-half of the trimmable fat tissue remaining on the cuts as purchased is actually ingested. If one were to exclude that fat tissue, the primary changes in dietary contribution would be in lipids and, of course, calories.

The challenges to red meat in the diet focus largely on its contribution to total dietary fat, saturated fatty acids, cholesterol, and sodium. Assuming that 2000 kcal represents the appropriate average energy intake for the total population, the calories from fat in red meat represent less than 12%, and only about 4.5%, or 25 kcal, of those 2000 kcal are supplied by saturated fatty acids.

For those who have been advised by a qualified health care professional to control dietary cholesterol, it is important to note that such an ingestion level supplies only about 90 mg of cholesterol daily. For those who are concerned about dietary sodium, fresh cooked red meats contain on the order of 60 to 80 mg per 85 g serving. In processed meats, which contain a higher level of sodium, sodium-containing compounds perform a number of essential functions for which viable alternatives do not now exist. Those functions include preservation, emulsion stabilization, myosin extraction to permit adhesion between pieces of meat, flavor enhancement, etc. More than 90% of the 528 mg of dietary sodium contributed by red meats is attributable to the processed meat component. For the approximately 80% of the population who need not be overly concerned about dietary sodium, that amounts to about 16–48% of the 1100–3300 mg range described as a safe and adequate range for adults (Food and Nutrition Board, 1980b).

Table 2
Daily Dietary Contribution of Red Meat
to the U.S. Diet, by User Group, 1984[a]

	Occasional/ Light		Moderate/ Light		Heavy		Total Population	
	Unit	%	Unit	%	Unit	%	Unit	%
Total meat								
g	41.14	—	117.00	—	216.31	—	110.985	—
oz	1.45	—	4.13	—	7.63	—	3.91	—
Protein (g)	9.47	16.9	26.33	47.0	47.57	84.9	25.62	45.7
Lipids (g)	8.48	—	24.46	—	45.55	—	23.84	—
kcal	117.5	5.9	334.9	16.7	618.1	30.9	326.4	16.3
Cholesterol (mg)	33.4	11.1	92.2	30.7	167.8	55.9	89.1	29.7
Iron (mg)	.91	9.1	2.49	24.9	4.44	44.4	2.37	23.7
Zinc (mg)	1.77	11.8	4.75	31.7	8.27	55.1	4.48	29.9
Sodium (mg)	160.9	3–15	526.4	16–48	1086.4	33–99	528.2	16–48
Thiamin (mg)	.091	6.5	.302	21.6	.605	43.2	.301	21.5
Riboflavin (mg)	.063	3.9	.189	11.8	.372	23.3	.189	11.8
Niacin (mg)	1.751	9.7	4.911	27.3	9.062	50.3	4.736	26.3
Vitamin B-12 (mcg)	.830	27.7	2.207	73.6	3.899	130.0	2.087	69.6

[a]Percentage of RDA for males ages 23–50 years (Food and Nutrition Board, 1980a); calorie percentage based on 2000 kcal mean energy needs of 154-lb sedentary U.S. adult male or active female. Sodium percentage based on 1100–3300 mg range of estimated safe and adequate daily intake for adults (Food and Nutrition Board, 1980b).

With regard to the positive contribution of red meat to the diet, it is widely recognized and generally accepted that red meat provides significant amounts of B vitamins, minerals, and nutritionally complete protein. Beef is recognized as a particularly rich source of iron, zinc, and vitamin B-12, while pork excels in thiamin content and is a very good source of zinc. All red meats are noted for their high nutrient density, which means that they supply a number of essential nutrients at much higher proportions of a human's requirements than their contribution to calorie needs.

Finally, good nutrition is achieved by satisfying each individual's dietary needs by appropriate selections from a variety of foods. In order to accomplish such satisfaction of human needs, it is essential that one have correct information regarding the nutritional profile of each food, as well as the amount of that food truly ingested. The intent of this document has been to clarify that dietary contribution for the red meat components of the diet as they are believed to be ingested.

Introduction

Historically, annual per capita red meat carcass production has carried the label "consumption." In an economic sense, that is a correct perception since it has all disappeared into use channels. In a nutritional sense, however, the term "food consumption" is generally perceived to be synonymous with "ingestion." Inasmuch as red meat carcasses are far from an ingestible form and much weight reduction must occur before they reach the stage of ingestible meat, it is apparent that the perceived interchangeability of those two terms has led to a significant overestimate of red meat ingestion.

Annual carcass weight disappearance includes literally billions of pounds of bones never to be eaten by humans, and also fat tissue, which is converted to other food uses or for other commercial purposes. Even retail cuts, as purchased, contain bones and trimmable fat, much of which is removed and becomes plate waste. Finally, ingested meat weight is reduced by cooking losses incurred in converting raw meat to a cooked state.

This paper has been prepared on the premise that a more realistic estimate of the nutritional status of Americans could be enhanced by the availability of more detailed data on the quantity of meat actually ingested. Since red meats, both nutritionally and economically, are of the utmost importance in the U.S. food supply, it is hoped that the data presented herein will be of use in a broad range of applications.

It is important to note at the outset that deriving meat ingestion

Table 3
Derived Weight by Class Breakdown of 1984 U.S. Beef FIS

	% Total Number of Head	Number of Head (Millions)	Average Carcass Weight (lb)	Total Carcass Weight (Billions of Pounds)	% Carcass Weight
Steers	46.5	16.6877	700	11.6814	51.7
Heifers	28.5	10.2113	615	6.2800	27.8
Cows	22.9	8.2285	495	4.0731	18.0
Bulls and stags	2.1	.7521	763	.5739	2.5
Total	100.0	35.8796	630	22.6084	100.0

figures and nutritional and physical composition figures requires the use of a number of estimates and assumptions. Published, objective information has been used to the extent that it was available, and estimates have been used only where deemed necessary. The overriding thrust of this document is to be certain not to underestimate either the total red meat ingestion or its accompanying lipids. Pursuant to that premise, for example, it is assumed that all the lean red meat tissue produced was ingested—no spoilage, plate discard of lean tissue, or other lean tissue loss occurred. In the matter of red meat contribution to dietary fat, necessary estimates were applied in such a fashion that fat contribution is believed to be nominally overstated. Because of the assumption of no lean meat losses through distribution and in the home and the assumption of no lean meat discarded (e.g., at restaurants), lean meat ingestion is overestimated. True red meat ingestion for both lean tissue and fat tissue is thus believed to be modestly overstated.

Table 4
Total 1984 U.S. Beef Slaughter

% Slaughter	Number of Head (Millions)	Slaughter Site
94.7	35.8796	FI
.8	.3100	Farm
4.5	1.7020	Other Commercial
100.0	37.8916	

Table 5
Deriving Net Available
Beef–Carcass Weight Basis[a]

Total slaughter weight	23.6956
Carcass condemnation	− .0996
Carcass shrink (.75%)	− .1770
Carcass disassembly (.25%)	− .0590
Net freezer stocks	− .0330
Exports and shipments	− .3830
Total	22.9440

[a]In billions of pounds.

Fresh Beef

The predominant component of the U.S. beef supply in 1984 was the domestic slaughter. Definitive breakdown by class and weight is available only for the federally inspected slaughter (FIS) portion (AMI, 1985; USDA, 1985) shown in Table 3.

Table 4 provides information about the total U.S. slaughter, including that identified as farm, federally inspected, and other commercial slaughter.

Adjusting for carcass condemnation, net freezer stocks, and exports and shipments (AMI, 1983), as well as carcass cooler shrink and disassembly loss (Weiss, 1985), results in a net reduction of the carcass weight figures by about .7516 billion lb (imports will be dealt with separately) to a net of 22.9440 billion lb (Table 5).

If one assumes that the remainder of the domestic slaughter is dispersed among the various classes in exactly the same proportions as in

Table 6
Estimated Dressed Weight Contribution
by Class to Total 1984 U.S. Slaughter

	Carcass Weight (Billions of Pounds)		%
	Total Slaughter	Adjusted to Table 5	Carcass Weight
Steers	12.2431	11.8548	51.7
Heifers	6.5820	6.3732	27.8
Cows	4.2689	4.1335	18.0
Bulls and stags	.6016	.5825	2.5
Total	23.6956	22.9440	100.0

Table 7
Estimated Compositional Profile by Class
of Total 1984 U.S. Slaughter

	% Carcass Weight			
	Steers	Heifers	Cows	Bulls and Stags
Lean tissue	54.9	52.9	56.4	75.2
Fat tissue				
Body cavity	2.7	3.5	3.0	1.0
Other	28.1	30.3	16.6	4.0
Bone	14.3	13.3	24.0	19.8
Total	100.0	100.0	100.0	100.0

the FIS, then the contribution to the carcass weight of each class can be calculated as in Table 6.

Figures are not available that describe the proportion of lean, fat, bone, and connective tissue in the American beef supply. However, the lean body mass of steers can be predicted (Breidenstein et al., 1968) if one estimates population characteristics widely used in deriving yield grade (USDA, 1980c) as follows: carcass weight 700 lb; fat thickness 0.6 in; kidney, pelvic, and heart fat 2.7%; and cross-sectional longissimus muscle area of 11.5 sq in at the 12th rib. That population of steers could be expected to yield 51.4% fat-free lean body mass. On the premise that lean trimmed of separable fat contains about 6.5% extractable fat, then the yield of edible lean tissue is about 54.9%. Using comparative industry yield figures, one can then estimate the lean tissue yield for each beef carcass class as in Table 7.

Using those composition figures and applying the carcass weights in

Table 8
Estimated Compositional Profile
for Total 1984 U.S. Beef Slaughter

	% Carcass Weight	Billions of Pounds
Lean tissue	55.1	12.6425
Fat tissue		
Body cavity	2.9	.6735
Other	26.1	5.9800
	15.8	3.6480
Total	99.9	22.9440

Table 9
Total Estimated Beef Tissue Available
for Use in the United States, 1984[a]

	Imports	Domestic	Total
Lean tissue	.8614	12.6425	13.5039
Fat tissue			
Body cavity		.6735	.6735
Other	.1852	5.9800	6.1652
Bone		3.6480	3.6480
Total	1.0466	22.9440	23.9906

[a]In billions of pounds.

Table 6, one can estimate the production of each of the tissues during 1984 (Table 8).

Imports of boneless beef in 1984 were 1.0466 billion lb (AMI, 1985, 1983). These imports are estimated by the authors of this document to consist of 82.3% lean tissue containing 3% extractable fat and 17.7% fat tissue containing 70.9% extractable fat for a composite 15.0% extractable fat. On those premises, the various beef tissues available for use in the United States can be computed (Table 9).

It is rational to assume that none of the bones or body cavity fats become a part of ingested red meat. Thus, total soft tissue that could become a part of consumable red meat consists of about 19.6691 billion lb.

The National Livestock and Meat Board (1982) estimates that 12% of the American beef supply is offered to consumers as processed meats. Beef tissue destined for processing thus is estimated at 2.3603 billion lb. No data are available to characterize that beef with regard to proportions of lean and fat tissue. In the absence of such characterization, the authors have estimated that beef processing raw material consists of 64.4% lean and 35.6% fat tissue containing 28 to 30% extractable fat for the composite (USDA, 1986) utilizing 1.5200 billion lb of lean tissue and .8403 billion lb of fat tissue.

The American Meat Institute (1985) estimates annual per capita production of hamburger to be 19.1 lb. The 1984 population of the United States is estimated by the U.S. Department of Commerce (1985) to be 236.1 million people. Total hamburger production for 1984 can thus be estimated at 4.5095 billion lb. There are again no data available to characterize ground beef raw material with regard to proportions of lean and fat tissue. The authors have, therefore, estimated that beef consists of 70.5% lean tissue and 29.5% fat tissue containing 25.4% extractable fat for that composite (USDA, 1986). Thus, the production of hamburger required 3.1781 billion lb of lean tissue and 1.3314 billion lb of fat tissue.

Table 10
Source of Beef Soft Tissue for "Cuts"

Source	Lean	Billions of Pounds Fat	Lean and Fat
Cows	.3497	.0847	.4344
Bulls and stags	.0621	.0033	.0654
Steers and heifers	8.3940	3.1656	11.5596
Total	8.8058	3.2536	12.0594

After accounting for edible lean tissue utilized for ground beef and processing use, a residual of 8.8058 billion lb of lean tissue remains to be used for other purposes. For purposes of this discussion, it is presumed to be used for fresh beef cuts. The authors have estimated that 15% of cow, bull, and stag lean tissue is ingested as "cuts" and, further that the soft tissue of cow "cuts" consists of 80% lean and 20% fat tissue; that "cuts" from bulls and stags consist of 95% lean and 5% fat tissue; and that "cuts" from steers and heifers consist of 72.7% lean and 27.3% fat tissue (USDA, 1986). It is estimated by the authors that about 28% of the chucks from fed cattle were used in either ground beef or processed meat. The soft tissue entities presumed to be sold as "cuts" thus are as shown in Table 10.

Table 11 indicates that we have a total of 6.1652 billion lb of fat tissue available for consumption produced with the red meat supply. The amount of fat used as a portion of the red meat supply is estimated at 5.4253 billion lb, leaving .7399 billion lb to be rendered, in addition to body cavity fat, for a total of 1.4134 billion lb rendered.

The authors estimate that 25% of the 4.5095 billion lb of ground beef produced in 1984 was used as an ingredient and the remaining 75% was

Table 11
Use Distribution of Beef Soft
Tissues[a]

Use	Lean	Fat	Total
Retail "cuts"	8.8058	3.2536	12.0594
Ground beef	3.1781	1.3314	4.5095
Processed meat	1.5200	.8403	2.3603
Rendered		.7399	.7399
Total	13.5039	6.1652	19.6691

[a]In billions of pounds.

grilled, pan-fried, or broiled. That 1.1274 billion lb used as an ingredient would presumably be similar in cooked composition to a baked product. Such use results in an overall cooked yield of 71% (USDA, 1986), or .8005 billion lb containing .1633 billion lb of lipids (20.4%). Since no values are available that describe loss or actual ingestion of rendered "cook out," the assumption was made that the losses were the same for fat and for moisture, although some consumers may, in fact, ingest some of the "cook out."

Ground beef remaining to be grilled, broiled, or pan-fried is estimated at 3.3821 billion lb raw and 2.3675 billion lb cooked, containing .4782 billion lb of lipids (20.2%). A cooked yield from the broiled or pan-fried ground beef composite can be derived as about 70% (USDA, 1986).

The beef sold in forms other than ground beef or processed meats is estimated at 12.0594 billion lb raw. One can derive a composited cooked yield of 69.0%, representing the mix of "cuts" as they existed (USDA, 1986), which reduces that raw weight to a cooked weight of about 8.3218 billion lb. As the composite is weighted for actual tissue available to be used as "cuts," the cooked boneless beef consisted of 72.73% separable lean tissue (6.0522 billion lb) and 27.27% separable fat tissue (2.2696 billion lb). The lean tissue was found to contain about 10.24% extractable lipid (.6197 billion lb). The cooked fat tissue is 70.3% extractable lipid, thereby providing 1.5955 billion lb of lipid.

The total contribution of fresh beef to dietary fat thus is a combination of ground beef and beef cuts, as shown in Table 12.

The amount of trimmable fat removed from cuts and discarded after cooking is highly speculative. The trim level applied by consumers to all meat amounts to roughly one-half of the separable fat on the cut (Rathje, 1985). It seems probable to the authors that estimating the removal of half such fat tissue would be conservative and would probably err on the

Table 12
Estimated Contribution of Fresh Beef
to Dietary Fat at Various Levels of Removal
of Trimmable Fat from Cuts[a]

| | Trimmable Fat Removed | | | | |
	0%	25%	50%	75%	100%
Ground beef	.6415	.6415	.6415	.6415	.6415
Separable lean	.6197	.6197	.6197	.6197	.6197
Separable fat	1.5955	1.1966	.7978	.3989	—
Total	2.8567	2.4578	2.0590	1.6601	1.2612

[a]In billions of pounds.

Table 13
Consumption Profile for Cooked Ground Beef

Consumer Category	Population Represented %	Number of People (Millions)	Annual Ingestion (Billions of Pounds)	Estimated Daily Per Capita Ingestion (Cooked) g	oz
Occasional users	10	23.610	.0325	1.71	.060
Moderate/light users	49	115.689	.9951	10.26	.362
Heavy users	38	89.718	2.1803	30.20	1.065
Total users	97	229.017	3.1629	17.19	.606
Non-users	3	7.083	—	—	—
Total population	100	236.100	3.1629	16.67	.588

high side in estimating lipid contribution of meat to the diet. On the premise that half the trimmable fat is removed and discarded, fresh beef contributes about 10.84 g of dietary fat per capita per day.

Ground beef is believed to be served in 97% of U.S. households and that in 87% of the households it is served at least once in a 2-week period (Yankelovich, Skelly and White, Inc., 1985). Households were divided into heavy users (serving ground beef twice a week or more) and moderate/light users (serving ground beef one to three times during a 2-week period). Of the two groups, heavy users were estimated to account for 43.7% of the users and 69% of the servings, whereas moderate/light users represented 56.3% of the users and 31% of the servings. In order to represent the entire user population, the authors of this document estimate that occasional ground beef users consume one-sixth as much ground beef per capita as the moderate/light user group. The estimated use profile is shown in Table 13.

The nutrient profile (USDA, 1986) derived for regular cooked ground beef was converted to a fat-free basis. Then, multiplying each such derived nutrient quantity by the derived lipid content of the Meat Board (MB) composited ground beef resulted in the nutrient content/100 g of the cooked MB composite. The nutritional contribution of ground beef to the diet can thus be calculated as shown in Tables 14 and 15.

It is believed that fresh beef other than ground beef is served in 99% of U.S. households and that it is served in 78% of the households at least once per 2-week period (Yankelovich, Skelly and White, 1985). The households were divided into use level groups, using the same criteria as for hamburger. Of the two groups, it was estimated that heavy users accounted for 32.1% of the users and 59% of the servings, whereas moderate/light users represented 67.9% of users and 41% of the serv-

Table 14

Daily Per Capita Contribution of Cooked Ground Beef to the U.S. Diet, 1984

Consumer Category	Protein (g)	Lipids (g)	kcal	Cholesterol (mg)	Iron (mg)	Zinc (mg)	Sodium (mg)	Thiamin (mg)	Riboflavin (mg)	Niacin (mg)	Vitamin B-12 (mcg)
Occasional users	.41	.35	4.9	1.5	.042	.088	1.3	.001	.003	.095	.048
Moderate/light users	2.46	2.08	29.2	9.2	.251	.528	8.0	.003	.019	.570	.287
Heavy users	7.24	6.11	86.0	27.1	.740	1.553	23.5	.009	.056	1.679	.846
Total users	4.12	3.48	49.0	17.2	.421	.884	13.4	.005	.032	.995	.481
Total population	4.00	3.37	47.5	15.0	.409	.857	13.0	.005	.031	.927	.467

Table 15
Percentages of RDA for Males Aged 23–50 Years Provided Per Capita by Cooked Ground Beef in the U.S. Diet, 1984[a]

Consumer Category	Protein	Lipids	kcal	Cholesterol	Iron	Zinc	Sodium	Thiamin	Riboflavin	Niacin	Vitamin B-12
Occasional users	.7	—	.25	.5	.42	.59	.04	.07	.19	.53	1.60
Moderate/light users	4.39	—	1.46	3.1	2.51	3.52	.24	.21	1.19	3.17	9.57
Heavy users	12.93	—	4.39	9.0	7.40	10.35	.71	.64	3.50	9.33	28.20
Total users	7.36	—	2.45	5.7	4.21	5.89	.41	.36	2.00	5.53	16.03
Total population	7.14	—	2.37	5.0	4.09	5.71	.39	.36	1.94	5.15	15.56

Source: Food and Nutrition Board, 1980a.

[a]Cholesterol percentage based on a daily intake of 300 mg dietary cholesterol per day. The USDA and USDHHS Dietary Guidelines (1986) say "Avoid too much fat, saturated fat, and cholesterol." Calorie percentage based on 2000 kcal mean energy needs of 154-lb sedentary U.S. adult male or active female. Sodium percentage based on 3300 mg, the top of the range of estimated safe and adequate daily intake for adults (Food and Nutrition Board, 1980b).

Table 16
Consumption Profile for Cooked Fresh Beef
Other than Ground Beef

Consumer Category	Population Represented		Annual Ingestion (Billions of Pounds)	Estimated Daily Per Capita Consumption (Cooked)	
	%	Number of People (Millions)		g	oz
Occasional users	21.0	49.581	.1835	4.60	.16
Moderate/light users	53.0	125.133	2.7774	27.58	.97
Heavy users	25.0	59.025	4.2261	88.98	3.14
Total users	99.0	233.739	7.1870	38.21	1.35
Non-users	1.0	2.337	—	—	—
Total population	100.0	236.076	7.1870	37.83	1.33

ings. In order to represent the entire user population, the authors of this document estimate the daily per capita use of the occasional users at one-sixth that of the moderate/light user group. The estimated profile of use is shown in Table 16, and dietary contributions in Tables 17 and 18.

Fresh Pork

United States pork production in 1984 was 14.8895 billion lb on a carcass weight basis (AMI, 1985), as shown in Table 19. Carcass condemnation, carcass shrinkage, cut floor losses, and changes in freezer stock decrease domestic pork availability to 14.3799 billion lb (Weiss, 1985). The lean and fat contribution among classes, as well as the contribution of imported pork, can be estimated from production numbers (Table 20). Pork primal cut yields as a percentage of carcass and their separable component compositional profile are shown in Table 21.

Based on these breakdowns among the classes, total soft tissue available for fresh or processed pork was 11.8426 billion lb.

The market direction of pork carcass components either to further processing or offered to the consumer as fresh pork items is not among the data collected and generally available. It is, however, common wisdom that most legs are further processed and most loins are sold to consumers in the fresh state. Spare ribs and neck bones are generally offered to consumers as fresh pork, whereas bellies, jowls, and trimmings are almost exclusively used for further processing. Estimated market direction and quantification estimates based on proportionalities of boneless, skinless meat mass (USDA, 1983) are presented in Table 22.

Table 17

Daily Per Capita Contribution of Cooked Fresh Beef Other than Ground Beef to the U.S. Diet, 1984

Consumer Category	Protein (g)	Lipids (g)	kcal	Cholesterol (mg)	Iron (mg)	Zinc (mg)	Sodium mg)	Thiamin (mg)	Riboflavin (mg)	Niacin (mg)	Vitamin B-12 (mcg)
Occasional users	1.25	.91	13.53	4.2	.129	.274	2.8	.002	.004	.173	.117
Moderate/light users	7.49	5.45	81.15	25.1	.774	1.645	16.9	.009	.024	1.039	.703
Heavy users	24.17	17.6	261.80	80.8	2.497	5.308	54.6	.030	.076	3.350	2.266
Total users	10.29	7.57	112.52	34.7	1.073	2.282	23.5	.013	.033	1.440	.974
Total population	10.19	7.47	111.22	34.3	1.060	2.255	23.3	.013	.032	1.423	.960

Table 18
Percentages of RDA for Males Aged 23–50 Years Provided Per Capita by Cooked Fresh Beef Other than Ground Beef in the U.S. Diet, 1984[a]

Consumer Category	Protein	Lipids	kcal	Cholesterol	Iron	Zinc	Sodium	Thiamin	Riboflavin	Niacin	Vitamin B-12
Occasional users	2.23	—	.68	1.40	1.29	1.83	.08	.14	.25	.96	3.90
Moderate/light users	13.38	—	4.06	8.37	7.74	10.97	.51	.64	1.50	5.77	23.43
Heavy users	43.16	—	13.09	26.93	24.97	35.39	1.65	2.14	4.75	18.61	75.53
Total users	18.38	—	5.63	11.57	10.73	15.21	.71	.93	2.06	8.00	32.47
Total population	18.20	—	5.56	11.43	10.60	15.03	.71	.93	2.00	7.91	32.00

[a]Cholesterol percentage based on a daily intake of 300 mg. The USDA and USDHHS Dietary Guidelines (1986) say, "Avoid too much fat, saturated fat, and cholesterol." Calorie percentage based on 2000 kcal mean energy needs of 154-lb sedentary U.S. adult male or active female. Sodium percentage based on 3300 mg, the top of the range of estimated safe and adequate daily intake for adults (Food and Nutrition Board, 1980b).

Table 19
Pork Carcass Weight Production
(Domestic) and Availability, 1984[a]

Total carcass weight available (domestic)	14.8895
USDA condemnation	− .0775
Carcass shrink (.75%)	− .1111
Carcass disassembly (.25%)	− .2940
Net freezer stock	− .0270
Total available carcass weight	14.3799

[a]In billions of pounds.

Based upon the assumption that matched cuts from the opposite sides of the same pork carcass are bilaterally symmetrical with regard to proportionality of lean and fat, one can use USDA data (1983) to estimate cooking losses of the lean and fat component of each of the primal cuts of pork. On that premise, and with the assumption that the cooking yield of the soft tissue of both spare ribs and neck bones is comparable to the weighted composite cooking yields of the lean/fat mass of the other primal cuts, one can then estimate cooked soft tissue destined for fresh consumption as shown in Table 23.

Separable lean and fat components of cooked primal cuts of pork (USDA, 1983) provide a partial basis for estimating fresh pork actually ingested. On the assumption that only one-half the separable fat on such cooked pork cuts is actually eaten and that the lean component is consumed in its entirety, then a rational estimate of pork actually ingested can be derived as shown in Table 24.

A nutritional profile of both cooked separable lean and separable fat (USDA, 1983) combined with the proportions of lean tissue and fat

Table 20
Total Fresh Pork Available for U.S. Consumption[a]

	Barrows and Gilts	Domestic Sows	Boars and Stags	Imported	Total
Lean tissue	5.9787	.4909	.1002	.2106	6.7804
Fat tissue	4.4169	.5143	.0581	.0728	5.0621
Bone/skin	2.5535	.2257	.0416	.0731	2.8939
Total	12.9491	1.2309	.1999	.3565	14.7364

[a]Carcass weight basis. In billions of pounds.

Table 21
Raw Pork Primal Cut Characterization

	Primals % of Carcass Weight		Edible Lean as % of Trimmed Primal	% Lipid in Edible Lean	Soft Tissue[c]	
	Regular[a]	Trimmed[b]			% Lean	% Fat
Leg	23.9	20.5	65.0	5.41	78.3	21.7
Loin	19.9	20.2	60.0	7.54	75.9	24.1
Blade (Boston)	7.3	7.1	70.0	9.28	79.5	20.5
Arm picnic	9.7	10.2	56.0	6.16	76.7	23.3
Spare ribs	3.9	2.9	62.0	23.6		
Neck bones	1.4	1.4	62.0	23.6		
Belly	16.8	13.9	31.0	9.28[d]		
50 L. trimmings	8.1	1.8	—			
Jowls	2.5	3.1	7.5			

[a]Average of USDA grades 1, 2, and 3—both belly and jowl reduced to 90% of reported yields to reflect removal of skin. [b]Cuts trimmed to levels comparable to those reported in USDA Handbook 8–10 (Rathje, 1985). [c]USDA Handbook 8–10. Neck bones are assumed to be the same as spare ribs in both yield of edible lean as percentage of primal and in soft tissue lipid content (Rathje, 1985). [d]Assumed to be comparable to Blade (Boston) lean.

Table 22
Estimated Market Direction of Boneless, Skinless Pork

	Soft Tissue from Primal Cuts (Billions of Pounds)	Used Fresh		Used Processed	
		% Fresh	Billions of Pounds	% Processed	Billions of Pounds
Leg	2.8627	8	.2290	92	2.6337
Loin	2.2870	85	1.9440	15	.3430
Blade (Boston)	.9536	30	.2861	70	.6675
Arm picnic	1.0989	5	.0550	95	1.0439
Spare ribs	.3420	100	.3420	—	—
Neck bones	.1228	100	.1228	—	—
Belly	2.2576	—		100	2.2576
Trimmings	1.1518	—		100	1.1518
Jowls	.3387	—		100	.3387
Fat tissue	.4275	—		100	.4275
Total	11.8426		2.9789		8.8637

Table 23
Estimated Cooked Fresh Pork

Primal Cut of Origin	Raw Weight (Billions of Pounds)	Lean and Fat Cooking Yield %	Cooked Weight (Billions of Pounds)
Leg	.2290	65.2	.1493
Loin	1.9440	73.0	1.4191
Blade (Boston)	.2861	62.0	.1774
Arm picnic	.0550	66.2	.0364
Spare ribs	.3420	67.4	.2305
Neck bones	.1228	67.4	.0828
Total	2.9789	71.6	2.0955

tissue as shown in Table 22 can thus be used to derive a nutrient profile of pork as believed actually to be ingested.

It would seem appropriate to state once again that the assumption is made that all lean tissue is actually eaten and that one can only speculate regarding the proportion of the separable fat tissue attached to the lean tissue of cooked retail cuts that actually becomes a part of dietary intake. Again, if one assumes that about one-half the fat tissue present on meat is eaten (Rathje, 1985), one would probably modestly overstate fat tissue consumption. Nonetheless, on that premise, the dietary contribution of fresh pork as eaten is portrayed in Table 25.

Fresh pork is believed to be served in 82% of U.S. households and is served in 54% of the households at least once per 2-week period (Yankelovich, Skelly and White, 1985). Households were divided into heavy users (serving fresh pork three times or more during a 2-week

Table 24
Estimated Cooked Fresh Pork Consumption

Primal Cut or Origin	Estimated % of Cook Meat Mass Eaten	Estimated Cooked Weight Consumed (Billions of Pounds)	Ingested % Contribution to Edible Portion Consumed	Pork % Lean	% Fat
Leg	94.0	.1403	7.1	91.9	8.1
Loin	93.5	1.3268	67.0	90.9	9.1
Blade (Boston)	93.5	.1659	8.4	91.9	8.1
Arm picnic	92.0	.0335	1.7	87.7	13.3
Spare ribs	100.0	.2305	11.6		
Neck bones	100.0	.0828	4.2		
Total		1.9798	100.0		

Table 25
Dietary Contribution Per 100 g Cooked Edible Fresh Pork

	Leg	Loin	Blade Boston	Arm Picnic	Neck Bones and Spare Ribs	Composite as Assumed To Be Eaten
Protein (g)	26.560	25.050	29.160	24.270	29.060	26.100
Lipids (g)	16.210	19.460	22.240	21.050	30.300	21.170
kcal	259.000	282.000	327.000	294.000	397.000	302.100
Cholesterol (mg)	94.000	90.000	114.000	95.000	121.000	97.100
Iron (mg)	1.060	1.110	1.910	1.290	1.850	1.290
Zinc (mg)	3.040	2.820	5.170	3.650	4.600	3.320
Sodium (mg)	61.500	65.700	71.600	74.600	93.000	70.300
Thiamin (mg)	.660	.757	.535	.550	.408	.674
Riboflavin (mg)	.330	.337	.365	.328	.382	.346
Niacin (mg)	4.740	5.660	4.190	4.120	5.480	5.420
Vitamin B-12 (mcg)	.710	.900	.880	.764	1.080	.911

period) and moderate/light users (serving fresh pork once or twice during a 2-week period). Of the two groups, heavy users accounted for 29.6% of the users but 55% of the servings, whereas moderate/light users accounted for 70.4% of users but 45% of the servings. In order to represent the entire user population, the authors of this document estimate that occasional users consume fresh pork at a level of one-sixth that of the moderate/light user group. The estimated profile of use is shown in Table 26.

Using data from Tables 25 and 26, one can generate the dietary contribution of fresh pork to the various user levels as shown in Table 27 and the relationship of that consumption to the recommended daily

Table 26
Consumption Profile for Cooked Fresh Pork

Consumer Category	Population Represented %	Number of People (Millions)	Actual Ingestion (Billions of Pounds)	Estimated Daily Per Capita Consumption g	oz
Occasional users	28.0	66.108	.1022	1.92	.068
Moderate/light users	38.0	89.718	.8331	11.54	.407
Heavy users	16.0	37.776	1.0445	34.36	1.212
Total users	82.0	193.602	1.9798	12.71	.448
Non-users	18.0	42.498	—	—	—
Total population	100.0	236.100	1.9798	10.42	.368

Table 27
Daily Per Capita Contribution of Cooked Fresh Pork to the U.S. Diet, 1984

Consumer Category	Protein (g)	Lipids (g)	kcal	Cholesterol (mg)	Iron (mg)	Zinc (mg)	Sodium (mg)	Thiamin (mg)	Riboflavin (mg)	Niacin (mg)	Vitamin B-12 (mcg)
Occasional users	.50	.41	5.8	1.9	.025	.064	1.4	.013	.007	.104	.017
Moderate/light users	3.01	2.44	34.9	11.2	.149	.383	8.1	.078	.040	.625	.105
Heavy users	8.97	7.27	103.8	33.4	.443	1.141	24.2	.232	.119	1.862	.313
Total users	3.32	2.69	38.4	12.3	.164	.422	8.9	.086	.044	.689	.116
Total population	2.72	2.21	31.5	10.1	.134	.346	7.3	.070	.036	.565	.095

Table 28
Percentages of RDA for Males Aged 23–50 Years Provided Per Capita by Cooked Fresh Pork in the U.S. Diet, 1984[a]

Consumer Category	Protein	Lipids	kcal	Cholesterol	Iron	Zinc	Sodium	Thiamin	Riboflavin	Niacin	Vitamin B-12
Occasional users	.89	—	.3	.63	.25	.45	.04	.93	.44	.57	.58
Moderate/light users	5.37	—	1.8	3.73	1.48	2.55	.24	5.55	2.49	3.47	3.50
Heavy users	16.01	—	5.2	11.12	4.43	7.60	.73	16.53	7.42	10.34	10.43
Total users	5.92	—	1.9	4.11	1.63	2.81	.27	6.11	2.74	3.82	3.85
Total population	4.85	—	1.6	3.37	1.34	2.30	.22	5.01	2.25	3.13	3.16

[a]Cholesterol percentage based on a daily intake of 300 mg. The USDA and USDHHS Dietary Guidelines (1986) say, "Avoid too much fat, saturated fat, and cholesterol." Calorie percentage based on 2000 kcal mean energy needs of 154-lb sedentary U.S. adult male or active female. Sodium percentage based on 3300 mg, the top of the range of estimated safe and adequate daily intake for adults (Food and Nutrition Board, 1980b).

Table 29
Estimated Compositional Profile
of Lamb and Mutton
in 1984 Slaughter[a]

	Lamb	Mutton
Edible lean tissue	52.0	52.9
Fat tissue		
Body cavity	5.6	3.1
Other	24.7	13.7
Bone	17.7	30.3
Total	100.0	100.0

[a]Percentage of carcass weight.

allowance (RDA) (Food and Nutrition Board, 1980a) for the 23- to 50-year-old male (Table 28).

Fresh Lamb and Veal

Lamb and mutton carcass weight available for use in 1984 was 0.3800 billion lb from domestic production and 0.0150 billion lb from imports (AMI, 1985, 1983). It is possible to estimate the body composition of lamb and mutton (Cole, 1966; Texas A&M University, 1986) shown in Table 29. Mutton is estimated to comprise 7.8% of lamb kill (American Sheep Producers' Council, 1986) or .0308 billion lb of the total kill (Table 30). Total soft tissue available for consumption from lamb and mutton in 1984 is presented in Table 31.

Table 30
Estimated Lamb and Mutton
Tissue Available for Use in the
United States, 1984[a]

	Lamb	Mutton	Total
Edible Lean tissue	.1894	.0163	.2057
Fat tissue			
Body cavity	.0204	.0010	.0214
Other	.0899	.0042	.0941
Bone	.0645	.0093	.0738
Total	.3642	.0308	.3950

Table 31
Total Lamb and Mutton Soft Tissue
Available for Consumption, 1984[a]

	Lamb	Mutton	Total
Lean	.1894	.0163	.2057
Other fat	.0899	.0042	.0941
Total	.2793	.0205	.2998

[a]In billions of pounds.

The lean tissue nutrient profile (Ono et al., 1984) is the basis for lamb dietary contribution; the fat tissue nutrient profile for beef is used (USDA, 1986) since no such data were available for lamb fat.

We assume that 15% of total soft tissue weight goes to processing (.0450 billion lb), with all mutton to processing (.0205 billion lb) and the remainder to processing from lambs (.0245). In the absence of good data, we assume that the lipid content of lamb/mutton going to processing is the same as for beef to processing (28–30% extractable lipid; 68.8% lean and 31.2% fat tissues). Therefore, .0310 billion lb of lean and .0140 billion lb of fat tissue were used for processed meats. The remaining soft tissue from lambs (.2548 billion lb) is comprised of .1747 billion lb of lean tissue and .0791 billion lb of fat tissue.

It is possible to derive a composited lean-to-fat ratio for lamb "cuts" if we assume that the ratio is similar to that for beef of 72.6% lean and 27.33% fat tissue containing 26.6% extractable lipid (USDA, 1986). It is assumed that all of the lean tissue (.1747 billion lb) and enough fat tissue is used to result in a composited ratio of 72.6% lean and 27.33% fat tissue requiring .0657 billion lb of fat tissue. Fat tissue remaining and not used in processed meat or lamb cuts is .0134 billion lb (Table 32).

When the appropriate cooking yields (Griffin et al., 1985) are compiled for the lean and fat portions of the cuts of lamb (70.3% overall), one

Table 32
Use Distribution of Lamb and
Mutton Soft Tissues, 1984[a]

Use	Lean	Fat	Total
Retail "cuts"	.1747	.0657	.2404
Processed meat	.0310	.0140	.0450
Rendered		.0134	.0134
Total	.2057	.0931	.2988

[a]In billions of pounds.

can derive a total lean contribution of .1228 billion lb cooked lean tissue. If it is again assumed that one-half of the separable fat is removed from the cut after cooking, the composite is derived containing 84.3% lean tissue and 15.7% fat tissue. Therefore, the fat tissue component of lamb "cuts" is .0229. The lean tissue nutrient profile (Ono *et al.*, 1984) composited for retail cuts was the basis for lamb dietary contribution. Since consumption use levels for lamb are not available as for other meats, per capita ingestion is expressed only on the basis of total population. The average person in the United States in 1984 consumed .77 g (.027 oz) cooked lamb per day.

The composited dietary contribution of lamb "cuts" is shown in Table 33 and the percent of the RDA for the 23- to 50-year-old male (Food and Nutrition Board, 1980a) provided per capita by fresh lamb in Table 34.

Veal available in the United States in 1984 was .4950 billion lb, produced domestically (AMI, 1985). Imported beef and veal are combined in the report and have been accounted for in the beef portion of this report. Veal carcass composition has been estimated, since industry figures for veal are not widely available.

Since use distribution data for veal is not available, the assumption was made that all veal tissue went to veal "cuts." While this assumption may overstate the contribution of fat from veal in the diet, soft tissue believed to go to "cuts" is comprised of .3340 billion lb lean and .0170 billion lb fat tissue. If we assume that the cooking yield of veal is somewhat lower than a lean, roasted beef product (owing to its lower total fat content), we can derive a total cooked yield of 69% for the composited lean and fat available from cuts. Therefore, the tissue available for ingestion was .2422 billion lb (.2305 billion lb lean and .0117 billion lb fat tissue).

Since consumption of veal was not defined by user group (Yankelovich, Skelly and White, Inc., 1985), equal distribution across the entire U.S. population is assumed. Therefore, veal ingested was 1.27 g (.045 oz) per capita per day in 1984. The nutrient composition of cooked veal cuts can be derived (Ono *et al.*, 1986) as shown in Table 35. The percentage of the RDA for the 23- to 50-year-old male (Food and Nutrition Board, 1980a) provided per capita by fresh veal is shown in Table 36.

Processed Meats in the Diet

In addressing the subject of the contribution of processed meats—including cured, smoked, and formulated meat products—to the U.S. diet, it may be appropriate to remind ourselves of the obvious; namely, their contribution to dietary variety in terms of flavor, texture, etc. Having stated the obvious, we must next recognize that processed meats face a number of significant challenges as a dietary component.

Table 33

Daily Per Capita Contribution of Fresh-Cooked Lamb to the U.S. Diet, 1984

Consumer Category	Protein (g)	Lipids (g)	kcal	Cholesterol (mg)	Iron (mg)	Zinc (mg)	Sodium (mg)	Thiamin (mg)	Riboflavin (mg)	Niacin (mg)	Vitamin B-12 (mcg)
Lamb	.200	.151	2.21	.79	.015	.037	.543	.0007	.0019	.0426	.0188

Table 34

Percentages of RDA for Males Aged 23–50 Years Provided Per Capita
by Fresh-Cooked Lamb in the U.S. Diet, 1984[a]

Consumer Category	Protein	Lipids	kcal	Cholesterol	Iron	Zinc	Sodium	Thiamin	Riboflavin	Niacin	Vitamin B-12
Lamb	.36	—	.15	.27	.148	.249	.016	.050	.118	.237	.6293

[a]Cholesterol percentage is based on a daily intake of 300 mg. The USDA and USDHHS Dietary Guidelines (1986) say, "Avoid too much fat, saturated fat and cholesterol." Calorie percentage based on 2000 kcal mean energy needs of 154-lb. sedentary U.S. adult male or active female. Sodium percentage based on 3300 mg, the top of the range of estimated safe and adequate daily intake for adults (Food and Nutrition Board, 1980b).

Table 35

Daily Per Capita Contribution of Fresh-Cooked Veal to the U.S. Diet, 1984

Consumer Category	Protein (g)	Lipids (g)	kcal	Cholesterol (mg)	Iron (mg)	Zinc (mg)	Sodium (mg)	Thiamin (mg)	Riboflavin (mg)	Niacin (mg)	Vitamin B-12 (mcg)
Veal	.413	.070	2.39	2.05	.018	.069	1.22	.0008	.0083	.1039	.0243

Table 36

Percentages of RDA for Males Aged 23–50 Years Provided Per Capita by Fresh-Cooked Veal in the U.S. Diet, 1984[a]

Consumer Category	Protein	Lipids	kcal	Cholesterol	Iron	Zinc	Sodium	Thiamin	Riboflavin	Niacin	Vitamin B-12
Veal	.74	—	.12	.68	.18	.46	.04	.06	.52	.58	.81

[a]Cholesterol percentage based on a daily intake of 300 mg. The USDA and USDHHS Dietary Guidelines (1986) say, "Avoid too much fat, saturated fat and cholesterol." Calorie percentage based on 2000 kcal mean energy needs of 154-lb. sedentary U.S. adult male or active female. Sodium percentage based on 3300 mg, the top of the range of estimated safe and adequate daily intake for adults (Food and Nutrition Board, 1980b).

Among those challenges are the oft-made criticisms that they are "fatty" foods. And, of course, the additives are viewed by some as having a potentially adverse effect on their value as a food entity. Among the additives, sodium currently appears to receive the most criticism.

Specific food consumption data for processed meats by the U.S. population are obviously nonexistent. As a result, any general statement regarding consumption must be recognized at the outset as an estimate. Reasonably good data frequently exist with regard to food production and its disappearance into consumption channels. In the case of meat products, however, as in the case of a number of foods, the production figures do not truly represent the form in which they are consumed. Rather, it may be in the form of carcass weight or retail cut weights estimated from carcass weights. Attempts are often made to further refine consumption data to represent raw weights of the consumable portion and still further to the consumable portion in a cooked form. Needless to say, a number of assumptions and estimates are necessary to convert production weights per capita to consumed food weights per capita.

It has been estimated that about 65% of pork, 12% of beef, and 15% of lamb soft tissues in the United States are converted to a processed form (National Livestock and Meat Board, 1982) before sale to the consumer. While such figures reflect the relative importance of processed versus fresh meats, they do not actually address the question of the contribution of processed meat to the diet.

Inasmuch as a number of assumptions are recognized to be necessary, the following is a statement of some of the key ones.

1. Edible portion consumption estimates are derivable from USDA Food Safety and Inspection Service (FSIS) production figures (AMI, 1985).
2. U.S. population estimates are suitably accurate (U.S. Department of Commerce, 1985).
3. In view of the absence of hard data, rational estimates can be made of:
   processing yields and proportionate uptake of processing solutions by the lean, fat, and bone components;
   lean, fat, and bone proportions in products as produced;
   edible portion yields through kitchen preparation; and
   nutrient composition of consumed edible portion.
4. The total production is consumed within the same year, with no provision for product losses through distribution and marketing.

The processed ham items pose a particularly difficult problem in deriving consumable product from produced weight. Each item requires a somewhat different set of assumptions. Following are the assumptions used in converting the various ham per capita weights to consumption weights.

Ham, bone in:

1. Edible portion consisting of 90% lean and 10% extractable fat constitutes 66.2% of uncured ham weight.
2. Yield through processing equals 100%.
3. Cooked yield of edible portion through kitchen preparation equals 90%.
4. Cooked edible portion equals 59.6% of the weight of the ham as produced for sale.

Ham, bone in, water added:

1. Edible portion consists of 91% lean and ingredients and 9% fat.
2. Yield through processing equals 112%, and 90% of the curing solution is taken up by the edible portion.
3. Cooked yield of edible portion through kitchen preparation equals 80%.
4. Cooked edible portion equals 55.9% of the weight of the ham as produced for sale.

Ham, semi-boneless:

1. Converting conventional uncured hams to a semi-boneless state results in the removal of the following:

Skin	6.4%
Fat	6.0%
Bone	5.2%
Trimmings	
(90% lean)	15.8%
Total removed	33.4%

2. Edible portion consists of 90% lean and 10% extractable fat and constitutes 75.7% of the weight of the ready-to-cure ham entity.
3. Yield through processing equals 100% of ready-to-cure ham weight.
4. Cooked edible portion yield through kitchen preparation equals 90%.
5. Cooked edible portion equals 68.1% of the weight of the ham as produced for sale.

Ham, semi-boneless, water added:
Assumptions 1 and 2 for ham, semi-boneless, apply.

3. Yield through processing equals 115% of ready-to-cure weight, and 90% of the curing solution is taken up by the edible portion.
4. Cooked edible portion yield through kitchen preparation equals 80%.
5. Cooked edible portion equals 62.05% of the weight of the ham as produced for sale.

Ham, boneless, and ham, sectioned and formed:

1. Processed ham entity contains 90% lean and 10% fat.
2. Cooked edible portion yield through kitchen preparation equals 90%.

Ham, boneless, water added, and ham, sectioned and formed, water added:

1. Processed ham entity contains 91% lean and ingredients and 9% fat.
2. Cooked edible portion yield through kitchen preparation equals 80%.

Ham, dry cured:

1. Edible portion consisting of 90% lean and 10% extractable fat constitutes 66.2% of uncured ham weight.
2. Yield through processing equals 80%.
3. Weight loss (water) from the edible portion equals 30.3% of the edible portion weight.
4. Cooked yield of edible portion through kitchen preparation equals 95%.
5. Cooked edible portion equals 54.9% of the weight of the ham as produced for sale.

Table 37 reflects the daily per capita weight contribution to the diet of hams processed under federal inspection, based upon previously defined potential consumers.

Other noncomminuted processed meats are grouped together solely for convenience. Following are the assumptions used in converting production weight of those products to per capita consumption weights:

Pork, regular:

1. Consists of a variety of bone-in and boneless cuts.
2. Edible portion consisting of 90% lean and 10% extractable fat represents 67.5% of the weight of the products as produced for sale.
3. Cooked yield of edible portion through kitchen preparation equals 90%.
4. Cooked edible portion equals 60.75% of the weight of the products as produced for sale.

Pork, water added:
Assumptions 1 and 2 from pork, regular, apply.

3. Cooked yield of edible portion through kitchen preparation equals 80%.
4. Cooked edible portion equals 54% of the weight of the products as produced for sale.

Bacon:

1. Cooked yield through kitchen preparation equals 31.9% (USDA, 1975b).

Table 37

Daily U.S. Per Capita Consumption of Smoked, Dried, or Cooked Hams Processed under Federal Inspection, 1984

	Production (g)	% Production Consumed	Consumption (g)
Bone in	.587	59.6	.350
Bone in, water added	1.714	55.9	.958
Semi-boneless	.060	68.1	.041
Semi-boneless, water added	.446	62.1	.277
Boneless	.494	90.0	.445
Boneless, water added	3.300	80.0	2.640
Sectioned and formed	.362	90.0	.326
Sectioned and formed, water added	1.811	80.0	1.448
Dry cured	.484	54.9	.266
Total hams	9.258	72.9	6.751

Table 38
Daily U.S. Per Capita Consumption of
Noncomminuted Processed Products Other than
Ham Processed under Federal Inspection, 1984

	Production (g)	% Production Consumed	Consumption (g)
Pork, regular	1.054	60.8	.641
Pork, water added	1.494	54.0	.807
Bacon	8.970	31.9	2.862
Cooked beef	1.792	100.0	1.792
Dried beef	.147	100.0	.147
Other meats	1.737	65.0	1.129
Total	15.194	48.6	7.378

Cooked beef and dried beef:

1. Assumed to be consumed as produced.

Other meats:

1. Consist of a variety of products.
2. Cooked edible portion equals 65% of the weight of the products as produced for sale.

Table 38 reflects the daily per capita weight contribution to the diet of noncomminuted smoked, dried, or cooked meat products (other than ham) processed under federal inspection, based upon previously defined potential consumers.

In the case of frankfurters, weiners, and bologna, they are assumed to be consumed as produced. Table 39 reflects the contribution of these products to the diet.

The assumption is made that all fresh and cured sausages will lose the same amount of weight through kitchen preparation as reported for fresh pork sausage (USDA, 1975b). Table 40 reflects the contribution of these products to the diet.

Dried and semi-dried sausages are presumed to be consumed as produced, and their contribution to the diet is reflected in Table 41.

Cured products are all assumed to have a yield through kitchen preparation as reported for corned beef (USDA, 1975a), and their contribution to the diet is reflected in Table 42.

The remainder of sausage-type products and canned products are

Table 39
Daily U.S. Per Capita Consumption
of Frankfurters, Weiners, and Bologna
Processed under Federal Inspection, 1984

	Consumption (g)		
	Franks and Weiners	Bologna	Total
Regular	6.100	2.712	8.812
With extenders	.585	.138	.723
With variety meats	.385	.434	.819
With extenders and variety meats	.326	.339	.665
Total	7.396	3.623	11.019

Table 40
Daily U.S. Per Capita Consumption
of Fresh and Cured Sausages Processed
under Federal Inspection, 1984

	Production (g)	% Production Consumed	Consumption (g)
Fresh beef	.059	55.9	.033
Fresh pork	4.623	56.5	2.612
Fresh, other	1.346	56.5	.760
Uncooked cured	.121	56.2	.068
Total	6.149	56.5	3.473

Table 41
Daily U.S. Per Capita
Consumption of Dried and Semi-
dried Sausages Produced under
Federal Inspection, 1984

Dried	1.401
Semi-dried	.442
Total	1.843

Table 42
Daily U.S. Per Capita Consumption
of Cured Products Produced under
Federal Inspection, 1984

	Production (g)	% Production Consumed	Consumption (g)
Cured beef	1.507	67.0	1.010
Cured, excluding beef and pork	.098	67.0	.066
Total	1.605	67.0	1.076

assumed to be 100% consumable as produced, and their contribution to the diet is reflected in Tables 43 and 44.

Table 45 reflects the weight contribution to the U.S. diet of the various groups of processed products and percentage contribution to total processed meat consumption of each product group, given all the assumptions previously stated. It is of interest to note that four clearly identifiable groups of formulated products constitute about 60% of the consumption of processed meats. Consumers of processed muscle cuts have the option of reducing dietary fat by removal of trimmable fat.

Such an option does not exist for comminuted products, although cooking can reduce the fat content of some of them. It is also important to recognize the efforts made on the part of meat processors to lower fat and sodium in processed products in order to better meet changing consumer demands.

Table 43
Daily U.S. Per Capita
Consumption of Other Meats
Processed under Federal
Inspection, 1984

	Consumption (g)
Liver sausage	.488
Cured meat loaves	.431
Nonspecific loaves	.743
Other cooked items	4.601
Other formulated products	1.388
Total	7.651

Table 44
Daily U.S. Per Capita
Consumption of Canned Meats
Processed under Federal
Inspection, 1984

	Consumption (g)
Canned hams	.817
Luncheon mats	.963
Pork picnics and loins	.089
Vienna sausage	.457
Franks and weiners	.004
Miscellaneous sausage products	.110
Deviled ham	.043
Sliced dried beef	.015
Chopped beef-hamburger	.066
Corned beef	.002
Total	2.566

Table 45
Daily U.S. Per Capita Consumption of Processed Meats, 1984

Domestic Production	Production (g)	% Production Consumed	Consumed (g)	% Consumed Processed Products
Smoked, dried, or cooked hams	9.258	72.9	6.751	15.34
Noncomminuted products other than ham	15.194	48.6	7.378	16.77
Franks, weiners, and bologna	11.019	100.0	11.019	25.05
Fresh and cured sausage	6.149	56.5	3.473	7.89
Dried and semi-dried sausage	1.843	100.0	1.843	4.19
Cured products	1.605	67.0	1.076	2.45
Other processed products	7.651	100.0	7.651	17.39
Canned products	2.566	100.0	2.566	5.83
Subtotal	55.29	75.5	41.757	94.91
Imported				
Canned hams and corned beef			1.561	3.55
Other			.676	1.54
Total			43.994	100.00

Table 46
Nutrient Information Sources
for Processed Meats

Smoked, dried, or cooked hams and pork	NDB 07029[a]
Smoked, dried, or cooked hams and pork, water added	NDB 07029 (:1.10)[a]
Hams, dry cured	Item 1767 (adjusted to 10% fat)[b]
Bacon	Item 1266[c]
Cooked beef	Item 355 (sodium estimated at protein × 3.5)[b]
Dried beef	Item 380[b]
Other meats	NDB 07029[a]
Frankfurters	NDB 07023[a]
Bologna	NDB 07008[b]
Fresh beef and other sausage	NDB 07065[a]
Fresh pork sausage	NDB 07064[a]
Uncooked cured sausage	NDB 07013[a]
Dried sausage	NDB 07072[a]
Semi-dried sausage	NDB 07078[a]
Cured products	Item 375[b]
Liver sausage	NDB 07041 (sodium estimated 1200 mg/100 g)[a]
Loaves and other formulated products	Average of NDB 07021, 07035, 07056, 07058, and 07062[a]
Other cooked items	Average of NDB 07059 and 07037[a]
Canned hams	Item 1783[b]
Canned luncheon meats	NDB 07045[a]
Canned Vienna sausage	NDB 07083[a]
All other canned meats	Average of item 1783, NDB 07045, and NDB 07083
Imported canned hams and shoulders	Item 1783[b]
Imported canned corned beef	Item 378 (sodium estimated at 1500 mg/100 g)[b]
Other import processed meats	Average of items 1783 and 378

[a]USDA, 1980a. [b]USDA, 1975a. [c]USDA, 1975b.

Inasmuch as nutrient composition data are not available for all categories of products offered for sale, additional assumptions must be made. Table 46 provides the USDA Nutrient Data Bank numbers (where available) that served as sources of nutrient content information on each group.

Table 47 is a composite of the estimated dietary components of processed meats. Although a great deal of compositional variability exists

Table 47
Dietary Components Per 100 g Processed Meats

Protein (g)	Lipids (g)	kcal	Cholesterol (mg)	Iron (mg)	Zinc (mg)	Sodium (mg)	Thiamin (mg)	Riboflavin (mg)	Niacin (mg)	Vitamin B-12 (mcg)
16.29	22.62	278	57	1.61	1.98	1077	.411	.211	3.624	1.15

Table 48

Estimated Consumption Profile for Processed Meats, 1984

Consumer Category	%	Population Represented Number of People (Millions)	Actual Ingestion (Billions of Pounds)	Estimated Daily Per Capita Consumption g	oz
Occasional users	26	61.447	.6681	13.51	.477
Moderate users	40	94.534	2.9253	38.46	1.357
Heavy users	25	58.869	4.7647	100.84	3.548
Total users	91	214.850	8.3581	48.35	1.705
Non-users	9	21.250			
Total population	100	236.100	8.3581	43.99	1.550

among the various products, the composite nutrient profile would appear to be useful.

In cases where referenced nutrient composition sources had no information for a dietary component, the average of the other products in that particular group was used to represent the total.

It is believed that processed meats were served in 91% of U.S. households during a 2-week period (Yankelovich, Skelly and White, 1985). Households were divided into heavy users serving processed meat 15 or more times during a 2-week period and light users serving processed meat 1 to 5 times during a 2-week period. Heavy users are estimated to account for 27.4% of the users and 57% of the servings, moderate users represented 44% of the users and 35% of the servings, and light users accounted for 28.6% of the users and 8% of the servings. The estimated use profile for processed meats is shown in Table 48.

Use of the composite processed meats nutrition profile of Table 47, applied to the estimated usage levels of Table 48, results in the derivation of the estimated daily dietary contribution of processed meat as shown in Tables 49 and 50.

Table 51 summarizes the estimated dietary contribution of the various groups of processed products in terms of protein, fat, kcal, and sodium. While a great deal of compositional variability exists among the various products, the total of the processed meat dietary components contains an estimated 16.3% protein, about 22.6% fat, about 278 kcal per 100 g, and about 1077 mg of sodium per 100 g. It seems plausible to assume that B-vitamins and minerals exist in about the same relationship to protein as is true for other meats.

As stated previously, hard consumption data simply do not exist for

Table 49

Daily Per Capita Contribution of Processed Meats to the U.S. Diet, 1984

Consumer Category	Protein (g)	Lipids (g)	kcal	Cholesterol (mg)	Iron (mg)	Zinc (mg)	Sodium (mg)	Thiamin (mg)	Riboflavin (mg)	Niacin (mg)	Vitamin B-12 (mcg)
Light users	2.24	3.19	39.5	8.0	.222	.272	148.4	.057	.029	.499	.158
Moderate users	6.38	9.06	112.4	22.7	.631	.774	422.0	.161	.083	1.419	.451
Heavy users	16.62	23.60	292.9	59.2	1.645	2.016	1099.5	.419	.216	3.698	1.174
Total users	8.01	11.38	141.2	28.5	.793	.972	529.9	.202	.104	1.782	.566
Total population	7.29	10.35	128.5	26.0	.721	.884	482.1	.184	.095	1.622	.515

Table 50
Percentages of RDA for Males Aged 23–50 Years Provided Per Capita by Processed Meats in the U.S. Diet, 1984[a]

Consumer Category	Protein	Lipids	kcal	Cholesterol	Iron	Zinc	Sodium	Thiamin	Riboflavin	Niacin	Vitamin B-12
Light users	4.01	—	1.97	2.7	2.22	1.81	5–13	4.04	1.82	2.77	5.28
Moderate users	11.39	—	5.62	7.6	6.31	5.16	13–38	11.49	5.18	7.88	15.02
Heavy users	29.68	—	14.64	19.7	16.45	13.44	33–100	29.95	13.48	20.54	39.12
Total users	14.31	—	7.06	9.5	7.93	6.48	16–48	14.44	6.50	9.90	18.86
Total population	13.02	—	6.42	8.7	7.21	5.89	15–44	13.14	5.91	9.01	17.16

[a]Cholesterol percentage is based on a daily intake of 300 mg. The USDA and USDHHS Dietary Guidelines (1986) say, "Avoid too much fat, saturated fat, and cholesterol." Calorie percentage based on 2000 kcal mean energy needs of 154-lb. sedentary U.S. adult male or active female. Sodium percentages based on 1100–3300 mg range of estimated safe and adequate daily intake for adults (Food and Nutrition Board, 1980b).

Table 51
Daily U.S. Per Capita Nutrient Contribution
of Processed Meat, 1984

	Consumption (g)	Protein (g)	Fat (g)	kcal[a]	Sodium (mg)
Smoked, dried, or cooked hams and pork	1.162	.204	.123	2.115	15.31
Smoked, dried, or cooked hams and pork, water added	5.323	.850	.511	8.807	63.73
Noncomminuted pork, regular and water added	1.448	.237	.143	2.456	17.77
Hams, dry cured	.266	.062	.026	.503	6.81
Bacon	2.862	.752	1.488	16.974	29.14
Cooked beef	1.792	.561	.109	3.587	1.91
Dried beef	.147	.050	.010	.300	6.33
Other meats	1.129	.198	.120	2.055	14.87
Frankfurters	7.396	.834	2.156	23.667	82.84
Bologna	3.623	.424	1.024	11.449	36.92
Fresh beef and other sausage	.793	.110	.199	3.141	6.39
Fresh pork sausage	2.612	.513	.814	9.639	33.80
Uncooked cured sausage	.068	.009	.017	.205	.38
Dried sausage	1.401	.321	.503	6.120	27.23
Semi-dried sausage	.442	.062	.089	1.105	4.70
Cured products	1.076	.246	.269	3.151	13.98
Liver sausage	.488	.069	.139	1.591	5.85
Loaves and other formulated products	2.562	.374	.340	5.175	34.05
Other cooked items	4.601	.630	1.286	14.631	44.91
Canned hams	.817	.149	.101	1.576	8.99
Canned luncheon meat	.963	.120	.292	3.217	12.41
Canned Vienna sausage	.457	.047	.115	1.275	4.36
All other domestic canned meat	.329	.044	.074	.884	3.66
Imported canned hams and shoulder	1.561	.285	.193	3.012	17.17
Imported canned corned beef and other processed meat	.676	.172	.060	1,258	9.82
Total	43.994	7.323	10.201	127.893	503.33

[a]Calorie figure represents combination of protein, fat, and carbohydrate calories (calculated).

foods. We are, therefore, forced to make a number of estimates and assumptions in order to interpret production data to per capita consumption. Composition for products produced, kitchen preparation losses, and plate waste are among those factors presumed to have the greatest potential impact on such an interpretation. Since no allowance has been made for spoilage or other product losses through distribution, the per capita consumption is believed to be somewhat of an overestimate of actual consumption.

Our daily per capita consumption of processed meats supplies 13.0% of an adult male's RDA for protein (Food and Nutrition Board, 1980a) and supplied only 4.8% of a 2000 kcal intake. The sodium content of the consumed processed meats is about 15% of the upper end and about ⅓ of the lower end of the range, depending on use level, of the safe and adequate range (Food and Nutrition Board, 1980b). If one is concerned about the percentage of dietary kcal originating from fat, processed meats as a group have about 73.3% of their energy originating from fat.

It would seem prudent to explore means by which both fat and sodium content could be reduced while still retaining the palatability characteristics and the current or superior food safety expectation for processed meats. It is common wisdom that we consume much more sodium than is metabolically required and that weight control is one of the most serious health problems in the United States.

Finally, processed meats have long added great taste appear variety to the diet. They are nutrient dense, meaning that their contribution to dietary protein, B-vitamins, and minerals is high compared to their energy contribution. As in all meats, the protein of processed meats is highly usable by the human, as is the iron and zinc that they provide. Processed meats have long been, and should continue to be, an important component of a palatable, balanced, and varied diet.

Conclusion

In using the consumption scenario presented herein, one must be cognizant of the limitations in terms of necessary assumptions and estimates and of dealing with averages. The limitations of averages apply both to food supply composition and to human consumption levels. Recognizing those limitations, it is believed that this presentation much more clearly reflects red meat actually ingested than is the case for "consumption" as carcass weight disappearance.

The estimated ingestion of each species (fresh), and of processed meat is summarized in Table 52. It seems unlikely that use levels of any meat product would be the same across all species. However, if one made that assumption, it would seem to come near to reflecting the minimum and maximum red meat contribution to the diet for red meat users.

For this definition of "heavy user," the high nutrient density of red meat is readily apparent. Such a user would derive about 37.9% of his

Table 52
Cooked Meat Ingestion for All Red Meats
in the Daily U.S. Diet, 1984

| | User levels | | | | | | Total Population | |
| | Light | | Moderate | | Heavy | | | |
	g	oz	g	oz	g	oz	g	oz
Fresh beef	16.31	.57	42.16	1.49	67.44	2.38	37.83	1.33
Ground beef	7.89	.28	17.50	.62	30.14	1.06	16.67	.59
Fresh pork	2.80	.10	11.17	.39	21.99	.78	10.42	.37
Fresh lamb	.32	.01	.62	.02	1.23	.04	.77	.03
Fresh veal	.63	.02	1.22	.04	2.41	.09	1.27	.04
Processed meat	13.18	.47	44.33	1.56	93.11	3.28	43.99	1.55
Total red meat ingested	41.13	1.45	117.00	4.12	216.32	7.63	110.95	3.91

2000-calorie dietary intake from meat, with less than 20% of those calories originating from saturated fat. That same red meat consumption would supply a much higher proportion of other essential nutrients, as shown in Table 53.

A more common way of viewing food ingestion is to spread the entire consumable amount that was produced and disappeared into consumption channels by the total population, also shown in Table 52. Of course, the same nutrient density is reflected in that such a level of use would provide about 16.3% of the calorie intake of that adult male with about 7% of his calorie intake originating from saturated fat. The cholesterol provided by that red meat intake represents less than 30% of the recommended limit of 300 mg per person per day.

For those who are concerned about dietary sodium, fresh meats generally range in the 60–80 mg level per 3-oz serving. In processed meats, however, sodium compounds perform a number of very important functions, including preservation, emulsion stabilization, and myosin extraction, to permit recombining meat chunks into a cohesive meat mass closely resembling natural whole muscle cuts, flavor, etc. About 91.3% of the sodium contribution of 528 mg originates with processed meats. For the vast majority of the population that need not be unduly concerned about dietary sodium, that 528 mg represents about 16% of the 3300 mg considered the upper end and 48% of the 1100 mg lower limit of the safe and adequate range for adults (Food and Nutrition Board, 1980b).

With regard to the positive contributions of red meat to the diet, it is

Table 53
Nutrient Contribution of Total Cooked Red Meat
Ingestion by Use Level in the U.S. Diet, 1984

Nutrient	User Levels			Total Ingested by All Users	
	Light	Moderate	Heavy	Units	% RDA[a]
Total meat (g)	41.140	117.000	216.310	118.890	
(oz)	1.450	4.130	7.630	4.190	
Protein (g)	9.470	26.330	47.570	26.440	47.2
Lipid (g)	8.480	24.460	45.550	24.950	—
kcal	117.500	334.900	618.100	340.100	17.0
Cholesterol (mg)	33.400	92.200	167.800	93.200	—
Iron (mg)	.910	2.490	4.440	2.500	25.0
Zinc (mg)	1.770	4.750	8.270	4.710	31.4
Sodium (mg)	160.900	526.400	1086.400	568.400	17.2–51.7
Thiamin (mg)	.091	.302	.605	.317	22.7
Riboflavin (mg)	.063	.189	.372	.198	12.4
Niacin (mg)	1.751	4.911	9.062	4.998	27.8
Vitamin B-12 (mcg)	.830	2.207	3.899	2.200	73.3

[a]Percentage of RDA for males ages 23–50 years (Food and Nutrition Board, 1980a); calorie percentage based on 2000 kcal mean energy needs of 154-lb. sedentary U.S. adult male or active female. Sodium percentage based on 1100–3300 mg range of estimated safe and adequate intake for adults (Food and Nutrition Board, 1980b).

widely recognized and accepted that red meat provides significant quantities of B-vitamins, trace minerals, and nutritionally complete protein. Beef is recognized as a particularly rich source of iron, zinc, and vitamin B-12, while pork excels in thiamin content and is also a rich source of zinc.

Good nutrition is a matter of satisfying each individual's dietary needs by appropriate selections from a variety of foods. In order to accomplish such satisfaction of human needs, it is essential that one have correct information regarding the nutritional profile of each food. The intent of this document is to clarify that nutritional profile for the red meat components in the diet as they are believed to be consumed.

As we very carefully define ingestion for all red meat, it is helpful to once again underscore the reality of "consumption" as it compares with "ingestion." A breakdown of carcass disappearance to consumer ingestion of red meat is provided in Table 54 for beef and pork. For beef, the 106.2 lb per capita disappearance converts to 43.8 lb fresh plus 7.6 lb processed beef ingestion per capita or 1.9 oz fresh and .33 oz processed beef per capita per day. Similarly, pork disappearance figures of 65.7 lb

Table 54

Converting Carcass Disappearance to Cooked Meat
Ingestion Per Capita in the United States, 1984

	Beef	Pork
Carcass disappearance (pounds per capita per year)[a]	106.20	65.70
Adjusted for carcass condemnation, freezer stock, import/freezer stock, import/export net, carcass shrink, and disassembly	101.60	62.40
Bones and kidney, pelvic and heart fat removed	83.30	50.20
Cooking losses and one-half separable fat removed[b]	43.80	8.50
Ingested/day, fresh (lb)	.12	.02
(g)	54.50	10.42
(oz)	1.92	.37
Ingested/day, processed (g)	9.21	34.60
(oz)	.33	1.22

[a]AMI, 1985. [b]12.0% of beef soft tissues and 74.9% of pork soft tissues are used in processing.

per capita become 8.5 lb fresh and 28.4 lb processed (.37 oz fresh plus 1.22 oz processed pork ingested per capita per day). Therefore, it would seem important to be clear on the actual term of ingestion or consumption being used in any discussion concerning the role of red meat in the diet.

References

American Sheep Producers' Council (1986). Personal Communication.

AMI (1983). Personal communication.

AMI (1985). "American Meat Institute Meatfacts." American Meat Institute, Washington, D.C.

Breidenstein, B. C., Lohman, T. G., and Norton, H. W. (1968). Comparison of the potassium-40 method with other methods of determining carcass lean muscle mass in steers. In "Body Composition in Animals and Man," Proc. Symp. National Research Council–National Academy of Sciences, Washington, D.C., pp. 393–412.

Cole, H. H. (1966). "Introduction of Livestock Production, Including Dairy and Poultry," 2nd ed. W. H. Freeman Co., San Francisco, California.

Food and Nutrition Board (1980a). "Recommended Dietary Allowances," revised. National Academy of Sciences–National Research Council, Washington, D.C.

Food and Nutrition Board (1980b). "Estimated Safe and Adequate Daily Dietary Intakes of Additional Selected Vitamins and Minerals." National Academy of Sciences–National Research Council, Washington, D.C.

Griffin, C., Savell, J., Smith, G., Rhee, K., and Johnson, H. K. (1985). Cooking times, cooking losses and energy used for cooking lamb roasts. *J. Food Quality* **8(2/3)**, 68.

National Livestock and Meat Board (1982). Internal working documents.

Ono, K., Berry, B. W., Johnson, H. K., Russek, E., Parker, C. F., Cahill, V. R., and Althouse, P. G. (1984). Nutrient composition of lamb of two age groups. *J. Food Sci.* **49(5)**, 1233.

Ono, K., Berry, B. W., and Douglass, L. W. (1986). Nutrient composition of some fresh and cooked retail cuts of veal. *J. Food Sci.* **51(5)**, 1352.

Rathje, W. (1985). Personal communication.

Texas A&M University (1986). "Nutrient Composition of U.S. and New Zealand Lamb." College Station, Texas.

USDA (1975a). "Composition of Foods: Raw, Processed, Prepared." Agriculture Handbook 8. Agricultural Research Service, U.S. Department of Agriculture, Washington, D.C.

USDA (1975b). "Nutritive Value of American Foods in Common Units." Agriculture Handbook 456. Agricultural Research Service, U.S. Department of Agriculture, Washington, D.C.

USDA (1980a). "Composition of Foods: Sausages and Luncheon Meats—Raw, Processed and Prepared." Agriculture Handbook 8-7. U.S. Department of Agriculture, Science and Education Administration, Washington, D.C.

USDA (1980b). Official United States standards for grades of carcass beef. U.S. Department of Agriculture, Agricultural Marketing Service, Washington, D.C.

USDA (1983). "Composition of Foods: Pork Products—Raw, Processed, Prepared." Agriculture Handbook 8-10. U.S. Department of Agriculture, Human Nutrition Information Service, Washington, D.C.

USDA (1985). "Agricultural Statistics." United States Government Printing Office, Washington, D.C.

USDA (1986). "Composition of Foods: Beef and Beef Products—Raw, Processed, Prepared." Agriculture Handbook 8-13. U.S. Department of Agriculture, Human Nutrition Information Service, Washington, D.C.

U.S. Department of Commerce, Bureau of the Census (1985). Population Estimates and Projections Series P-25, 922, Washington, D.C.

USDA/USDHHS (1986). "Nutrition and Your Health: Dietary Guidelines for American." Home and Garden Bulletin 232-3. U.S. Department of Agriculture, Human Nutrition Information Service, Washington, D.C.

Weiss, G. (1985). Internal working documents. National Livestock and Meat Board, Chicago, Illinois.

Yankelovich, Skelly and White, Inc. (1985). Consumer climate for meat products. Prepared for American Meat Institute, Washington, D.C., and National Livestock and Meat Board, Chicago, Illinois.

Demographics of Dairy Cows and Products

Paul A. Putnam

U.S. Department of Agriculture
Agricultural Research Service
Stoneville, Mississippi

Table 1
Milk: Supply and Utilization, United States, 1930–1985[a]

	1930	1935	1940	1945	1950	1955	1960	1965	1970	1975	1980	1985
Milk production by cows on farms	100,158	101,421	109,510	121,504	116,602	122,945	123,109	124,180	116,962	115,398	128,406	143,147
Utilization (Milk equivalent)												
Butter	32,162	32,665	36,801	27,285	28,641	28,951	29,374	28,458	23,925	19,873	22,826	24,385
Cheese												
American	3,904	4,813	6,115	8,777	8,972	10,073	9,686	11,548	14,266	16,452	23,765	28,556
Other	1,157	1,424	1,747	2,346	2,883	3,480	3,678	4,277	5,300	7,431	10,160	13,144
Cottage	—	—	—	—	—	765	960	1,046	1,248	1,028	1,016	917
Canned												
Evaporated	3,113	3,947	5,266	8,147	6,320	5,490	4,342	3,471	2,686	2,020	1,594	1,509
Condensed	715	447	614	816	622	734	972	931	579	686	524	817
Dry whole milk	118	156	223	1,650	952	822	721	654	507	464	607	877
Frozen (ice cream, etc.)	3,602	2,973	4,712	5,788	7,808	9,651	11,306	12,557	13,078	14,046	14,010	15,021
Total manufactured	54,764	57,019	62,679	61,357	55,039	58,027	59,751	61,768	59,992	60,524	73,327	85,537
Total fluid	32,066	30,564	33,519	46,000	43,309	49,200	53,000	55,400	52,000	51,123	50,855	52,014
Total on farm	11,207	12,646	12,072	11,671	9,808	14,625	9,158	5,974	4,002	3,061	2,338	2,456

[a]In millions of pounds.

Table 2
Pounds Per Capita U.S. Civilian Consumption
of Selected Dairy Products, 1910–1985

Year	Fluid Milk and Cream	Condensed and Evaporated Milk	Dry Whole Milk	Nonfat Dry Milk	Butter	Cheese	Ice Cream
1910	315	5.8	<0.1	—	18.1	4.2	1.9^a
1915	318	9.4	<0.1	—	17.0	4.1	3.8
1920	348	8.5	<0.1	—	15.0	4.0^a	7.5
1925	337	11.5	<0.1	0.4	17.8	4.6	9.6
1930	337	13.4	0.1	1.3	17.3	4.6	9.6
1935	326	16.0	0.1	1.6	18.3^b	5.2	8.0
1940	331	19.1	0.1	2.2	16.7	5.9	11.2
1945	399^b	18.0	0.4	1.9	10.9	6.6	15.4
1950	349	19.8^b	0.3	3.6	10.6	7.6	17.0
1955	352	16.2	0.3	5.5	9.0	7.9	18.0
1960	322	13.7	0.3	6.2	7.5	8.3	18.3
1965	302	10.6	0.3	5.6	6.4	9.6	18.5
1970	264	7.1	0.2	5.4	5.3	11.5	17.7
1975	243	5.3	<0.1	3.3	4.7	14.3	18.5^b
1980	227	3.8	0.3	3.0	4.5^a	17.6	17.3
1985	220^a	3.7^a	0.4	2.2	4.9	22.6^b	18.0

^{a}Low value. ^{b}High value.

Introduction

Statistics may not be for everyone, but they do tell a lot about what has happened to us and how we live. In this case, statistics tell what has happened to the dairy industry and how our dairy product consumption patterns have changed.[1]

First, to set the stage, Table 1 shows what has happened to the milk production and utilization in the United States since 1930. Total milk production has increased by over 40%. Cheese production has increased by 800%. Ice cream and other frozen products have increased by 400%. Manufactured dairy products and fluid milk produced have increased by 55 and 65%, respectively.

On the other hand, butter production has decreased by 25% and milk consumed on the farm has decreased by 80%. These decreases are presumably due to the availability of the good quality lower priced spreads and health considerations for the former and because of the more than 80% reduction in the farm population for the latter.

---

1. The main reference for this chapter is U.S. Department of Agriculture (1949–1985), *Agricultural Statistics 1949–1985*, Washington, D.C.

Table 3
Milk: Milk Cows[a] in Specified Countries, 1950–1985

	1950	1955	1960	1965	1970	1975	1980	1985	% Change
Australia	2,354	2,338	3,244	3,214	2,677	2,356	1,869	1,804	−23
Austria	1,185	1,178	1,127	1,110	1,085	1,026	975	995	−16
Belgium	950	985	1,008	1,007	1,033	1,061	1,040	1,031	+9
Canada	3,609	3,312	3,162	2,885	2,471	2,135	1,727	1,723	−52
Denmark	1,577	1,483	1,438	1,350	1,153	1,094	1,032	896	−43
Finland	—	1,155	1,153	1,138	—	—	—	628	−46
France	8,400	9,370	9,812	8,470	9,039	10,207	7,452	6,764	−19
Germany, West	5,602	5,749	5,797	5,816	5,593	5,405	5,469	5,547	−1
Ireland	—	1,198	1,279	1,547	1,669	1,230	1,449	1,549	+29
Italy	—	—	3,700	3,387	3,500	2,882	3,013	3,174	−14
Japan	—	—	—	—	1,060	910	1,070	1,101	+4
Netherlands	1,518	1,510	1,599	1,698	1,920	2,210	2,340	2,354	+55
New Zealand	1,846	1,995	1,989	2,032	2,363	2,080	1,999	2,165	+17
Norway	766	658	603	517	421	387	372	381	−50
Sweden	1,681	1,514	1,299	994	746	675	656	646	−62
Switzerland	858	886	940	920	901	884	875	816	−5
United Kingdom	3,767	3,706	4,013	4,205	4,471	3,220	3,392	3,311	−12
United States	21,944	21,193	17,543	14,954	12,000	11,151	10,810	11,025	−50

[a]1000 head.

In the same period (1910–1985) that the above changes occurred, the U.S. population increased by 260%.

The impact of these changes is reflected in pounds per capita consumption, summarized in Table 2. Fluid milk and cream reached a peak consumption of 399 lb in 1945 and decreased to 220 lb per capita by 1985. Similarly, condensed milk peaked in consumption in 1950 at 19.8 lb and decreased to 3.7 lb per capita in 1985.

Dry whole milk has not been a major consumption item, and nonfat dry milk has been only a little more successful as a dairy product. The decreased production of butter is even more strikingly evident in its decrease in consumption from over 18 lb in 1935 to less than 5 lb per capita in 1985.

Where, then, is all the milk going? The rising stars of the dairy industry in the United States have been cheese and ice cream. Consumption of both have risen steadily since 1910. It looks as if ice cream is plateauing at about 18 lb per capita, but cheese seems to still be on an upward trend at over 22 lb consumption per capita in 1985.

On the international level, there have been similarly important trends. As in the United States, dairy cow numbers have decreased

Table 4

Milk: Yield[a] Per Cow in Specified Countries, 1950–1985

	1950	1955	1960	1965	1970	1975	1980	1985	% Change
Australia	2.481	2.737	1.963	2.172	2.887	2.831	2.976	3.473	+ 40
Austria	1.670	2.155	2.526	2.903	3.069	3.158	3.483	3.779	+123
Belgium	3.408	3.766	3.869	3.921	3.690	3.644	3.734	3.957	+ 16
Canada	2.072	2.374	2.653	2.893	3.369	3.627	4.550	4.580	+121
Denmark	3.439	3.462	3.762	3.984	4.024	4.495	4.958	5.691	+ 65
Finland	—	2.478	3.037	3.381	—	—	—	4.909	+ 98
France	1.844	1.965	2.314	2.875	3.020	2.908	3.751	3.967	+115
Germany, West	2.479	2.946	3.328	3.650	3.916	3.997	4.531	4.628	+ 87
Ireland	—	2.099	2.295	2.030	2.175	3.683	3.331	3.904	+ 86
Italy	—	—	2.613	3.006	2.677	3.183	3.568	3.222	+ 23
Japan	—	—	—	—	4.503	5.454	6.077	6.701	+ 49
Netherlands	3.810	3.863	4.284	4.215	4.300	4.625	5.036	5.331	+ 40
New Zealand	2.564	2.546	2.679	2.939	2.505	2.841	3.416	3.638	+ 42
Norway	2.092	2.459	2.680	3.195	4.202	4.682	5.218	5.178	+148
Sweden	2.916	2.727	3.028	3.685	3.968	4.693	5.282	5.720	+ 96
Switzerland	2.946	3.152	3.287	3.371	3.461	3.818	4.177	4.712	+ 60
United Kingdom	2.783	2.922	3.164	2.713	2.777	4.068	4.705	4.935	+ 77
United States	2.415	2.641	3.182	3.775	4.430	4.697	5.393	5.911	+145

[a]In metric ton.

Table 5
Milk: Total Production[a] in Specified Countries, 1950–1985

	1950	1955	1960	1965	1970	1975	1980	1985	% change
Australia	5,840	6,400	6,368	6,984	7,728	6,670	5,562	6,265	+ 7
Austria	1,979	2,538	2,847	3,216	3,330	3,253	3,411	3,760	+ 90
Belgium	3,238	3,710	3,908	3,950	3,812	3,869	3,898	4,080	+ 26
Canada	7,477	7,863	8,402	8,346	8,324	7,744	7,858	7,891	+ 6
Denmark	5,423	5,135	5,410	5,378	4,640	4,918	5,117	5,099	− 6
Finland	—	2,862	3,502	3,772	—	—	—	3,083	+ 8
France	15,490	18,366	22,700	24,351	27,302	30,910	29,450	26,830	+ 73
Germany, West	13,889	16,942	19,290	21,228	21,902	21,604	24,778	25,674	+ 85
Ireland	—	2,514	2,935	3,141	3,630	3,683	4,826	6,047	+141
Italy	—	—	10,409	9,782	9,370	9,760	11,423	10,277	− 1
Japan	—	—	—	—	4,773	4,963	6,502	7,378	+ 55
Netherlands	5,783	5,716	6,852	7,157	8,256	10,221	12,005	12,550	+117
New Zealand	4,735	5,079	5,329	5,971	5,920	5,909	6,828	7,683	+ 62
Norway	1,602	1,640	1,638	1,679	1,809	1,834	1,967	1,975	+ 23
Sweden	4,904	4,128	3,933	3,663	2,960	3,168	3,465	3,695	− 25
Switzerland	2,579	2,831	3,119	3,124	3,140	3,396	3,679	3,845	+ 49
United Kingdom	10,483	10,829	12,697	11,410	12,415	13,100	15,958	16,340	+ 56
United States	53,455	56,013	55,820	56,442	53,165	52,371	58,298	65,166	+ 22

[a]In metric tons.

Table 6
Milk: Production in Specified Countries, 1985

	Milk Cows (1000 Head)	Yield Per Cow (Metric Ton)	Total Milk Product (1000 Metric Ton)
Argentina	2,950	1.974	5,823
Brazil	14,700	0.707	10,400
Chile	660	1.576	1,040
China	1,200	2.083	2,499
Czechoslovakia	1,830	3.760	6,883
Finland	628	4.909	3,083
Germany, East	2,080	4.348	9,044
Greece	335	1.825	648
Hungary	624	4.634	2,723
India	27,700	0.686	19,000
Mexico	5,087	1.360	6,920
Peru	690	0.935	622
Poland	5,528	3.000	16,530
Portugal	253	3.154	740
Romania	2,550	1.402	3,700
South Africa	1,885	1.234	2,327
Soviet Union	43,600	2.262	98,608
Spain	1,910	3.298	6,330
Venezuela	1,345	1.218	1,638
Yugoslavia	2,640	1.772	4,679

since 1930 in most of the countries of the world listed in Table 3. Exceptions are The Netherlands, Ireland, New Zealand, and Belgium, where numbers of cows increased by 56, 29, 8, and 5%, respectively. The increases amounted to about 1.2 million head as compared to a decrease of nearly 11 million head in the United States alone. The net change among the countries listed in Table 3 approximates a reduction of 20 million milk cows. It should be noted, however, that a very significant number of cattle occurs in countries not listed in Tables 3, 4, and 5. Additional countries are listed for 1985 in Table 6. Information from these countries were not included in earlier source publications or were excluded by the author because of other reasons. Especially to be noted are the large numbers of cows in Brazil, India, and the Soviet Union.

Table 4 provides data on yield per cow; in all countries listed, yield has increased substantially. In Austria, Canada, France, Norway, Sweden, and the United States, yield per cow has about doubled or more since 1950. Yields per cow in Czechoslovakia, Finland, East Germany, Hungary, Poland, Portugal, and Spain for 1985 (Table 6) equal or approach yields of cows in the specified countries listed in Table 4. Earlier

Table 7
Butter: Production in Specified Countries, 1950–1985[a]

	1950	1955	1960	1965	1970	1975	1980	1985	% Change
Argentina	46	57	59	44	28	40	29	32	− 30
Australia	—	205	196	205	215	161	84	114	− 44
Austria	—	—	—	43	46	45	43	43	=
Belgium	58	91	87	81	88	101	97	105	+ 81
Brazil	41	50	50	50	53	65	90	70	+ 71
Canada	150	154	151	157	150	129	111	108	− 28
Czechoslovakia	—	—	55	82	87	109	128	150	+173
Denmark	145	165	167	166	132	139	113	110	− 24
Finland	34	60	37	102	87	74	74	73	+115
France	185	293	293	446	513	556	627	595	+222
Germany, East	—	—	159	173	216	270	280	316	+ 99
Germany, West	247	327	380	502	506	521	578	515	+109
Greece	4	10	10	13	8	7	—	5	+ 25
Hungary	—	—	15	17	20	18	32	31	+106
India	—	—	—	—	—	451	480	700	+ 55
Ireland	52	59	65	67	72	86	124	163	+213
Italy	48	65	65	66	53	62	76	72	+ 50
Japan	—	—	—	—	45	40	63	89	+ 98
Netherlands	71	74	85	104	121	204	179	229	+223
New Zealand	—	200	217	253	222	240	255	293	+ 47
Norway	15	15	82	20	19	20	22	25	+ 66
Poland	—	—	84	105	128	251	294	308	+266
Portugal	1	2	3	2	4	2	4	4	+300
Romania	—	—	9	19	31	37	—	47	+422
South Africa	26	39	43	39	45	26	17	17	− 35
Soviet Union	—	—	808	1186	1069	1231	1373	1596	+ 98
Spain	—	—	—	—	7	13	22	15	+114
Sweden	103	88	85	80	50	56	66	74	− 29
Switzerland	17	27	31	35	29	34	35	37	+118
United Kingdom	19	23	31	39	68	48	165	202	+963
United States	725	705	675	617	520	445	519	585	− 19
Yugoslavia	—	10	21	—	18	16	18	10	=

[a] In 1000 metric tons.

yield data was generally not available from the Agricultural Statistics references cited for most of these countries.

Even though numbers of dairy cows have decreased dramatically in the majority of the countries listed, increased production per cow has more than offset the effect of the diminished cow populations. The Soviet Union and the United States are by far the largest producers of total milk products. In fact, these two countries produce about as much milk products as the rest of the countries in the world combined!

Table 8
Cheese: Production in Specified Countries, 1950–1985[a]

	1950	1955	1960	1965	1970	1975	1980	1985	% Change
Argentina	93	130	120	139	162	226	245	220	+ 136
Australia	45	40	43	59	86	99	142	160	+ 255
Austria	7	16	22	34	46	60	74	83	+1086
Belgium	6	11	19	37	39	41	24	37	+ 517
Brazil	42	60	60	70	110	147	240	205	+ 388
Canada	54	40	50	79	98	121	177	213	+ 294
Czechoslovakia	—	—	37	50	64	90	109	131	+ 254
Denmark	57	87	104	115	111	147	221	253	+ 344
Finland	8	22	26	38	41	56	70	79	+ 888
France	183	300	389	528	637	943	1,145	1,300	+ 610
Germany, East	—	—	32	41	127	184	210	246	+ 669
Germany, West	127	159	155	151	273	618	421	495	+ 290
Greece	39	72	77	101	130	157	168	193	+ 395
Hungary	—	—	10	16	25	23	38	51	+ 410
India	—	—	—	—	—	—	—	—	—
Ireland	2	3	3	16	33	62	49	78	+3800
Italy	223	349	373	421	396	529	615	684	+ 207
Japan	—	—	—	—	—	9	9	20	+ 122
Netherlands	106	174	187	220	284	375	443	522	+ 392
New Zealand	99	97	96	105	104	89	106	118	+ 19
Norway	20	30	36	45	52	62	70	72	+ 260
Poland	—	—	17	24	44	283	94	118	+ 653
Portugal	1	2	2	4	7	24	37	39	+3800
Romania	—	—	32	55	70	75	—	97	+ 203
South Africa	8	13	13	15	20	22	31	34	+ 325
Soviet Union	—	—	—	—	479	559	687	809	+ 69
Spain	—	—	—	—	21	100	140	101	+ 381
Sweden	52	54	52	59	60	81	96	109	+ 110
Switzerland	47	60	64	78	84	104	125	126	+ 168
United Kingdom	32	64	103	115	131	239	237	260	+ 713
United States	525	620	641	792	1,000	1,275	1,807	2,279	+ 334
Yugoslavia	—	—	—	—	—	120	147	52	− 57

[a]In 1000 metric tons.

Although butter production has decreased by nearly 20% in the United States since 1950 and decreased by greater percentages in Argentina, Australia, Canada, Denmark, South Africa, and Sweden, total world production of butter has increased (Table 7). Especially notable increases in butter production have occurred in France, East and West Germany, The Netherlands, Poland, the Soviet Union, and the United Kingdom.

Cheese production stands in a class by itself as a dairy product. All countries, except Yugoslavia, have increased cheese production dramatically (Table 8). Nearly all increases are several hundredfold since 1950!

Statistics do not document why these changes have occurred, but certainly demand is a part of the reason for the shift to cheese. Other factors such as concerns about health and animal fat consumption, storage and marketability, and life-styles are probably contributory.

Production changes relate to the development and recovery of certain countries, better management (breeding, feeding) of dairies, and response to consumer demands.

Whatever the case, it is clear that dairy has been a growth industry since the turn of the century, especially since the conclusion of World War II (1950 and after).

Future

Projections on Future Productive Efficiency[1]

L. W. Smith[2]

U.S. Department of Agriculture
Agricultural Research Service
Beltsville Agricultural Research Center
Animal Science Institute
Beltsville, Maryland

1. Originally published in "Future Agricultural Technology and Resource Conservation." Chap. 10, Red Meat, Dairy, Poultry, and Fish Technology, Work Group Report. Iowa State University Press, Ames, Iowa, 1984.
2. A. A. Andersen, R. Crom, H. E. Dietz, J. P. Fontenot, D. E. Friedly, R. J. Gerrits, W. Hansel, B. G. Harmon, R. A. Michieli, R. R. Oltjen, I. T. Omtvedt, D. B. Polk, J. M. Shelton, and R. W. Touchberry helped contribute to this chapter.

Handbook of Animal Science

Introduction

A strong animal agriculture is essential to resource conservation and is complementary with good conservation practices. Improved range forage production–ruminant utilization systems have great potential for increased production of high-quality protein while providing an opportunity to achieve greater resource conservation.

Animal products supply 53% of all food consumed in the United States, including 69% of the protein, 40% of the energy, 80% of the calcium, 67% of the phosphorus, 36% of the iron, and significant percentages of vitamins and trace minerals of our diets. These statistics support the importance of animals to our food supply. Protein of animal origin is highest of all sources in biological value. In the United States only 16% of our disposable income is spent on food, the lowest of all nations. There is a continued and growing demand for animal protein in the developing countries. Animal numbers need not change in order to meet growing demands, as this may be met through improved productivity. The number of breeding animals may in fact decrease in the years ahead. The per capita consumption of animal products in the United States is expected to remain unchanged, but shifts from one source to another are anticipated. The feeding of grain to ruminants is expected to continue for the foreseeable future. Grain feeding is not synonymous with producing fat beef. Continued improvement in animal production efficiency is expected. Improved animal production efficiency is anticipated through advances in use of superior germ plasm, application of genetic engineering to meet animal health and production challenges, improved reproductive efficiency, and innovative uses of more forages and underutilized feedstuffs.

Projections presented in Tables 1, 2, and 3 are results of the collective judgment of the authors. Projections on poultry were made by a group of poultry specialists assembled by the first author after the conference.

Table 1
Potential for Increased Efficiency of Producing Foods
of Animal Origin in 2000 and 2030[a]

Animal/Product	Unit	% Improvement	
		2000	2030
Beef	Live weight marketed per breeding female	25	60
Pork	Live weight marketed per breeding female	35	60
Dairy	Milk marketed per breeding female	30	65
Sheep	Live weight marketed per breeding female	35	70
Broiler chickens	Live weight marketed per breeding female	30	35
Turkeys	Live weight marketed per breeding female	40	40
Laying hens	Number of eggs	20	25
Fish (catfish)	Age to market weight (1 lb)	50	200

[a]Best estimate of authors.

Projections on Future Productive Efficiency for Products of Animal Origin

The projections for increased efficiency in producing foods of animal origin are shown in Table 1. A CAST (Council of Agricultural Science and Technology) report (1980) presented historical efficiency data on

Table 2
Food Protein Production Efficiency by Animals[a]

Food Product	Rate of Production	Production Efficiency of Protein (g Per Mcal DE)	% Improvement	
			2000	2030
Eggs	233 eggs/yr	12.6	10	15
Broiler	1.6 kg/8.0 weeks	15.9	15	20
Turkey	10 kg/18 weeks 2.2 kg feed/1 kg gain	22.0	20	30
Pork	.504 kg gain/day	6.1	12	25
Milk	13.7 kg/day	12.4	10	20
Lamb	.22 kg/day	2.6	15	30
Beef	1.05 kg gain/day	2.3	15	25

[a]Adapted from Reid et al., 1980. Production efficiency data include maintenance cost. Best estimate of authors. Provided by the authors.

Table 3
Per Capita Consumption in 2000 and 2030[a]

Item	1990 Retail Weight	1990 Carcass Weight	2000 Retail Weight	2000 Carcass Weight	2030 Retail Weight	2030 Carcass Weight
Beef	77	104	80	108	80	108
Veal	2	2	2	2	2	2
Pork	60	65	60	65	60	65
Lamb	2	3	2	3	2	3
Dairy products	504		520		510	
Chicken						
Broilers	49		55		55	
Mature chickens	3		3		3	
Turkey	11		13		15	
Eggs	36		33		30	
Fish	13		18		27	

[a]Pounds per person.

farm animals for the years 1925, 1950, and 1975. The work group noted that swine had made considerable improvement in efficiency between 1975 and 1982.

Beef

Assumptions used in improving efficiency for beef production include marketing young intact males, greater use of superior germ plasm, improved reproductive efficiency, use of new feed additives, and use of more forages and underutilized materials in innovative ways.

It is expected that protein and fat deposition will be controlled by manipulating the kind of nutrients presented to sites of absorption and manipulating the animal's endocrine system. Genetic engineering will lead to improved disease control through new vaccines and disease-resistant animals. Genetic engineering also is expected to provide new ruminant microorganisms capable of breaking the lignin–cellulose bond, fixing nitrogen in the rumen, and producing nutrients of the quality and quantity needed for particular productive functions.

Pork

In the swine industry, pigs marketed per sow per year are expected to increase from the current level of 13.8 of 1982 to 18 in 2000 to 20 in 2030.

These increases in pigs marketed per sow and an increase in market weight from 240 lb in 1982 to 250 lb in 2000 and 265 lb in 2030 will result in live weights marketed per breeding female of 3312, 4500, and 5300. Improved efficiency will be achieved by introduction of new and superior germ plasm, improved reproductive efficiency, and control of the rate of deposition of protein and fat. Along with improved genetic ability, improvements in nutrition will be needed to meet the demand of sows producing large litters and to meet the higher nutrient requirements associated with increased rates of gain.

Dairy

It is expected that average milk production in the United States will increase from the present 12,300 lb per cow per year at the present to approximately 16,000 lb by the year 2000 and to 20,300 lb by the year 2030; these are increases of 30 and 65%, respectively. Average milk production per cow increased approximately 200% from 1945 to 1982, with approximately half of that increase coming in the period 1966–1982, when the increase was approximately 3800 lb. For the period 1982–2000 it is anticipated that average milk production per cow will increase by approximately 3800 lb of milk and 223 lb of milk fat per cow per year. This increase will result from the excellent genetic evaluation program that is in place for bulls and cows and which should continue to improve during the next 20 years. There may not be as much gain resulting from a shift from small, less specialized to larger, more specialized dairy farms in the coming 18 years as in the past 18, but many of the smaller 50 to 100 cow dairy herds will adopt more effective feeding, reproductive, health care, and milk management programs to keep the rate of increase essentially linear.

For the period 2000–2030 average milk production per cow is expected to increase by 4300 lb. The potential for increase is considerably higher and will depend on the rate of adoption of new technology such as increased use of proven sires and dams; use of embryo transfer; the use of products of recombinant DNA such as hormones, vaccines, and other biologicals; embryo manipulation procedures; and continually improving technology for feeding, milking, reproductive management, and health care. If U.S. cows averaged 20,000 lb of milk per cow per year, 120 billion lb could be produced with 6 million cows; with an average of 25,000 lb of milk per cow only 4.2 million dairy cows would be required. If cow numbers are reduced drastically, serious social and economic problems would exist in important dairy states such as Wisconsin, Minnesota, New York, and Pennsylvania.

Sheep

Sheep numbers have declined in the United States for several decades, but this trend appears to have leveled out or reversed in recent years. Sheep are extremely well adapted to conservation agriculture. This applied to both range area and to marginal farmlands throughout the country that might be converted to grass or forage production as contrasted to those uses that contribute to soil and water loss. Additionally, sheep can produce saleable products (lambs and wool) without benefit of extensive grain feeding. They also produce red meats with equal or greater efficiency than competing ruminant species, and perhaps even more efficiently than the monogastric species when only harvested grains or concentrate feeds are considered. There is a very great potential for improved productive efficiency from sheep through improved reproductive rate resulting from frequent lambing and a larger number of lambs per lambing. Genotypes recently have been developed or introduced into this country that should accomplish these goals in the near future. If this potential is realized in commercial flocks, sheep should show the greatest response in productive efficiency of any species of domestic livestock. Two problems facing the sheep industry are public acceptance of lamb at the marketplace and seasonability of product. This may be solved partially through population increases, but the greatest potential is through innovative market development.

Goats

Reliable statistics on goats in the United States are not available, but the number is not likely to exceed 5 million head. On a world basis the number exceeds 400 million. Goat numbers in the United States are increasing and are almost certain to continue in the future. Three types of goats are found in this country—dairy goats kept for milk production, the Angora kept primarily for mohair production, and meat type goats. Angora goat numbers are influenced greatly by demand for mohair and as a result have fluctuated over the years. Numbers of milk or meat type goats are expected to increase. The potential for goat production is very high, but for the foreseeable future goats are not expected to become a significant factor in the U.S. agricultural picture. Goats are useful in biological control of brush and forbs and can produce food and fiber from otherwise unusable forage resources. Since goats will forage selectively on many invading and troublesome browse species they can reduce the need for the use of herbicides with the attendant environmental concern or disruptive mechanical control practices. Another potential advantage for goats is their suitability for practicing conservation agriculture on small acreage holdings. All three types of goats are expected to respond to selection and to other technological developments in the future.

Poultry

It is expected that poultry production efficiency will continue to increase, as shown in Tables 1 and 2. Assumptions used in arriving at the percentage of improvement for broilers and turkeys include increased eggs laid per hen, improved hatchability, and rearing to a heavier market weight to accommodate further mechanical processing. Superior germ plasm will allow improved rate of gain, and genetic engineering will contribute new vaccines leading to improved disease resistance and reduced mortality. Improved efficiencies in egg production will occur through a small improvement in rate of lay with the major impact from introduction of dwarf germ plasm and forced molting. These factors will lead to reduced feed per dozen eggs.

Rabbits

Rabbits are excellent convertors of feed to meat. Present-day good doe production can give 8 litters of 64 offspring weighing 4 lb each, resulting in 256 lb per doe per year. Demonstrated feed efficiencies are 2.7 feed to 1 gain. While present consumption in the United States is less than .5 lb per capita, consumption in Europe is 4–5 lb.

Catfish

Catfish production efficiency is expected to increase based on reducing age to market weight of 1 lb from 18 mo for 1982 to 12 mo in 2000 and to 6 mo in the year 2030. These advances will occur as a result of long-term genetic improvement and adaptation or developing tolerance to low levels of dissolved oxygen. The improvement by the year 2030 will happen with the aid of intensive culture systems using elevated water temperature and control systems that minimize pollutant accumulation. The work group noted other types of fish, and shellfish are also grown for food.

Other Considerations

Protein production efficiency by animals in rate of production per unit time and protein production per Mcal DE is shown in Table 2. Data of

Reid *et al.* (1980) for what they termed approximately average U.S. management conditions were used in projecting improvement in Table 2. Analysts are cautioned to take note of units of expression of production efficiency data and to understand fully limitations and implications of conversions to other units. Scientific advances anticipated and mentioned specifically for one species will also contribute to progress in production efficiency of others. Continued application of computer technology is expected to make further contribution to efficiency of animal agriculture as a whole.

Per Capita Consumption of Foods from Animal Origin

Per Capita Meat Consumption

All meat products are consumed at a market clearing price in the short run. In the longer run, 20 to 50 years hence, production can be changed to respond to consumer demand. Even beef production can be adjusted in this time frame. Population growth will lead to an increase in total production and consumption of all meat, poultry, and fish.

On a carcass-weight equivalent basis, 275 lb per person is 0.75 lb per day (about 0.65 lb retail weight). Since a large proportion of the population does not consume meat at breakfast, 0.75 lb per day allows 0.25 lb at lunch and 0.5 lb at the evening meal. Both represent servings for the average person.

In 1982, per capita consumption of red meat, poultry, and fish was 230 lb retail weight (265 carcass weight), and per capita consumption may be reaching a "saturation" point—the capacity of an individual's daily intake. Consumption of the 3 minor meats (veal, lamb, and mature chicken) is expected to hold constant for the projection period.

Overall red meat, poultry, and fish consumption is projected at 233 lb and 244 lb retail weight for the years 2000 and 2030. This is 267 lb and 278 lb when beef and pork are converted to carcass equivalent. Thus, by 2030 consumption would equal 0.76 lb carcass weight (0.67 retail weight) per day.

Consumption projections by species indicate a significant uptrend for fish, broiler, and turkey, a modest uptrend for beef, and no change in pork, lamb, and goat consumption. While population will lead to increases in total consumption, consumption of dairy products and eggs per person is expected to decline slowly (.1–.3% per year) over the next 50 years. The rate of decline probably will slow after the year 2000.

Regional Shifts in Animal Production

There is no major shift or rapid change seen in regional animal production in the United States, only gradual adjustment over time.

Hogs

The north–central states have long accounted for about 80% and the southeastern states about 12–15% of total production. Over the past 30 years the southeast has gained slightly in the percentage of total production. The southeast has climatic advantages but is faced with an increasing feed cost disadvantage (chiefly grain) due largely to higher costs of transportation. The southeast probably has reached its maximum in terms of total hog production. Hog production has been shifting east to west (slowly within the north–central region. Hog production likely will continue to concentrate in the western Corn Belt and eastern parts of the Northern Plains, where grain prices are lowest and soybean meal is plentiful.

Cattle Raising

Breeding beef cattle are distributed nationwide in areas of cheap forage supplies. Growth of beef production in the southeast to 25% of U.S. supply occurred in 1950 and later was related closely to cheap nitrogen needed for high forage yields. Unless problems from legume diseases can be resolved, the southeast probably will have a lessening share of the U.S. total of beef production because of high-cost nitrogen, soybean production, and conversion of much pastureland to timber production. The central United States, especially the western part of the north–central states, have the strongest potential for expansion. Costs and returns are likely to remain stable for many years so that traditional range areas won't be able to afford costly range renovation to expand production beyond ceiling levels at which they now operate.

Cattle Feeding

Continued concentrations will remain in the band stretching from the High Plains to Iowa, with Minnesota, Kansas, and Nebraska defining the corners of the area. As the water supply drops, grain production will lessen and production will become relatively more costly in the southern part of this area. There will be some shift to the northeast portion of the area. Specialized cattle feeding will remain in the west to the extent of availability of special feeds, especially various residues.

Dairy

No particular shifts are foreseen in present dairy production regions.

Poultry

No regional shifts are expected in production of poultry.

Sheep

If sheep production expands, it will more than likely do so in the crop farming areas under intensive systems of production and not in the range areas where presently concentrated.

Issues that Relate to Animal Agriculture

The group identified several regulations and conflicts that may reduce the rate of increase in advancing productivity. These issues were not discussed.

Impact of government regulation on production and conservation
Animal welfare
Predator control
Brush control—herbicides—pesticides—insecticides
Land use and lay-aside programs administered by different
    agencies
Water-use policies
Air pollution—related to feed lots
Waste disposal—stream pollution-odor
Certification of feed mills
Registration for feed additives
EPA regulations on registration and use of pesticides
Public lands grazing policies that inhibit adoption of new produc-
    tion technologies
Public lands policies that discourage red meat production
Registration of pesticides and new drugs
Federal agencies in U.S. Department of Agriculture deemphasiz-
    ing programs related to maintaining or increasing production
    of red meat, food, and fiber
Conflicting programs in agriculture between various agencies
Farm programs that have adverse effects on conservation from
    resource leasing

Disease control and eradication program
Veterinary services on vaccines
Conflicting research reports and articles and action based on them
about nutritional and health issues

One of the lesser-recognized changes that has become important in the meat industry recently is product structuring. This structuring has allowed the industry to take portions of the carcass and restructure it into a uniform, highly desirable serving size unit. The success of chicken and pork sandwiches and chicken "nuggets," available in fast-food outlets, may be creating both an additional market and a more highly valued product. This change could increase the level of product consumption and certainly will increase the percentage of carcass for human food.

Research Needs

Advances in animal production efficiency through genetic engineering, cloning, manipulating endocrine systems to advantage, manipulating protein–fat deposition through nutrition and other means, developing resistance to diseases, and other advances too numerous to list here will contribute to increased production efficiency. Research with animals costs twice as much as other kinds of agricultural research. This cost is associated with labor, feed, and facilities to maintain farm animals and high costs of hormones, drugs, and other biological materials. For continued success in achieving increased animal production efficiency a few obvious areas need intensified research attention. Ruminant(s) and swine production efficiency can be improved greatly by increasing the number of offspring produced per year. Genetically superior animals exist today but insufficient knowledge exists about nutritional requirements for these animals. For these animals to express their genetic potential, nutritional requirements need to be matched to genetic potential. More research is needed on understanding how to change and control the deposition of protein and fat by endocrinological, nutritional, and possibly other methods. Lignocellulose is the most abundant material on earth. It is a major constituent of diets of ruminants. Discovery of how to economically break the bond between lignin and cellulose to totally free cellulose as an energy source for ruminant microorganisms would be an accomplishment of major significance. Pond (1980) presents more detailed descriptions of research imperatives in animal agriculture.

References

CAST (1980). Foods from animals: Quantity, quality and safety. CAST Report No. 82. Council of Agricultural Science and Technology, Ames, Iowa.

Pond, W. G., Merbel, R. A., McGilliard, L. D., and Rhodes, V. J. (1980). "Animal Agriculture: Research to Meet Human Needs in the 21st Century." Westview Press, Boulder, Colorado.

Reid, J. T., White, Ottilie D., Anrique, R., and Fortin, A. (1980). Nutritional energetics of livestock: Some present boundaries of knowledge and future research needs. *J. Anim. Sci.* **51,** 1393.

Acknowledgments

The work group acknowledges the assistance of D. D. King, National Program Staff, ARS, USDA; T. J. Sexton, Avian Physiology Laboratory, ARS, USDA, Beltsville, Maryland; and R. W. Rosebrough, Nonruminant Animal Nutrition Laboratory, ARS, USDA, Beltsville, Maryland, in projecting improved efficiencies for poultry in this report.

Index